CELESTIAL DELIGHTS

Celestial Delights

The Best Astronomical Events through 2001

Francis Reddy
and
Greg Walz-Chojnacki

CELESTIALARTS

Berkeley. California

*To my parents, Elizabeth and Francis, who introduced
me to the Cape Cod sky.*

FR

*To my mother — a woman whose love has been a guiding star to all her
children — who just missed a dedication the last time around.*

GWC

Text and cover design by David Charlsen
Typesetting by ImageComp

Cover: Figure 1-1. *The crescent moon and the bright planet Venus set over Tulsa,
Oklahoma, in April, 1988. Note how the separation between moon and planet changes
in this multiple-exposure sequence (start at the top). The moon moves so swiftly that its
eastward orbital motion carries it noticeably farther from Venus by the time the pair
sets. (Photo by Bill Sterne)*

FIRST CELESTIAL ARTS PRINTING 1992

Library of Congress Cataloging in Publication Data

Reddy, Francis, 1959–
    Celestial delights : the best astronomical events through 2001 / Francis Reddy and Greg
Walz-Chojnacki.
        p.   cm.
Includes bibliographical references and index.
ISBN 0-89087-675-4
    1. Astronomy—Observers' manuals.   2. Astronomy—Amateurs' manuals.   I. Walz-
Chojnacki, Greg, 1954– II. Title.
QB63.R33   1992
520—dc20   92-27724
                                           CIP

1 2 3 4 5 6 7 8 9 10 / 96 95 94 93 92

CONTENTS

PREFACE

Astronomy has a bad reputation.

Everyone takes a moment to sky-gaze now and then — admiring the colors of a sunset, searching for the Man in the Moon, or playing connect-the-dots with the Big Dipper. There is, nevertheless, a widespread impression that the only folks who can really enjoy this age-old hobby are those blessed with dark country skies, expensive equipment, and lots of time on their hands.

That's far from the truth. The skies under which most of us live, while dimmed by the orange glow of city lights, are more than adequate for observing the phenomena that fascinated and puzzled astronomers during the first two thousand years of human civilization: the motions and arrangements of the sun, moon, and planets. Several times a year, planets, stars, and the moon gather in striking arrangements. You'll need no equipment to appreciate this beguiling and often beautiful dance of the moon and planets.

Our aim in this book is to share with you the simple beauties of the sky that our ancestors admired. You'll see the motions that captivated the ancients with your own eyes, which will in turn reveal to your mind's eye the grand pattern of the planets first perceived by Copernicus. By using the moon and brighter planets as celestial signposts, the tables and charts in this book guide you to every planet that can be seen with the unaided eye. Other chapters detail upcoming eclipses, introduce the major constellations, and describe the best meteor showers. The Appendix contains an almanac of easy-to-see astronomical events from 1993 through 2001 — eclipses, meteor showers, planetary gatherings, and moon phases all organized by date and time and cross-referenced with the diagrams and tables elsewhere in the book. So you can use *Celestial Delights* both as an introduction to astronomy and as a calendar of celestial events.

We hope you will come to enjoy the beauty of the heavens that is everyone's heritage. We encourage you to participate in this centuries-old delight. All we ask is that you occasionally look up and wonder "What is that bright star?"

Acknowledgments

We wish to thank all of the people who contributed to this book, both directly or indirectly, but especially: astronomer Fred Espenak of NASA's Goddard Space Flight Center, for walking us through the nuances of eclipses and for providing predictions for our maps in Chapter 3 (which are largely derived from his own published work); our friend Robert Miller, for helpful discussions and some much-needed custom computer programs; Ken Crossen of TechView Corp. in Carrboro, North Carolina, for introducing us to the technical drawing program MASS-11 DRAW; and Nancy Mack, founding editor of *Odyssey* magazine, whose encouragement and support sustained us both for many years.

INTRODUCTION

We live in an age when the complex and forbidding explanations of science often mask the simple beauty of nature. The wonder of a thunderstorm lies hidden behind satellite photos, weather charts, and barometric pressures. The astounding fact of human conception and birth seems lost amidst ultrasounds, amniocentesis, and the lingo of genetic engineering. The grandeur of astronomy, the study of our very place in the universe, likewise appeals to our sense of wonder and our need to explore our origins. But discuss a celestial event with even an amateur astronomer, and you'll soon find yourself swimming in terms like "retrograde motion," "arc-seconds," and "right ascension."

Now, scientific inquiry has its own kind of beauty, and special terms (okay, jargon) become a useful form of shorthand — you'll learn a few yourself in this book. The problem is we're often led to feel that the scientific appreciation of nature must be left wholly to those who have had years of formal training, or who devote a large part of their free time to science as a hobby. The good news is that isn't true — and it's especially untrue of astronomy. As citizens of the twentieth century, a time in which humans have walked on the moon and robot spacecraft have explored other worlds, we're all somewhat familiar with the layout of the solar system and the basic laws that govern the universe.

Knowing that the planets circle the sun is one thing, but recognizing how that movement expresses itself in the sky above us is quite another. The reason that most of us don't appreciate the heavens is simply that we haven't taken the time to look. The ancients may not have *understood* the sky as well as we do, but they certainly *knew* it better. Yet we have the advantage. The basic motions that fascinated the ancients remain on display. We have the opportunity to both know *and* understand. Obviously you're willing to look — this book is in your hands, right? — so the sky is within your grasp.

The Meaning of the Sky

The two celestial objects we are most familiar with are naturally the brightest: the sun and moon. The sun has always been recognized as vitally important, and was viewed by many ancient cultures as the greatest deity. Even the most primitive peoples appreciated the life-giving powers of what we now regard as our star. The reverence shown to the sun was reflected by the astronomical functions adopted by ancient priests. These first astronomers came to recognize the annual variations in the sun's pattern of rising and setting. They used them to mark the march of the seasons and guide essential activities, such as planting and harvesting.

Even today, many of our holidays and holy days reflect the seasons. For example, scholars tell us Christ was born in the spring, yet Christmas falls near the winter solstice, that hopeful date when the days begin to lengthen, promising greater warmth and sunshine to come. The present timing of Christmas is in part the result of an attempt to supplant pagan celebrations, such as the Roman Saturnalia and the Teutonic Yule festival, with a Christian one. Easter and

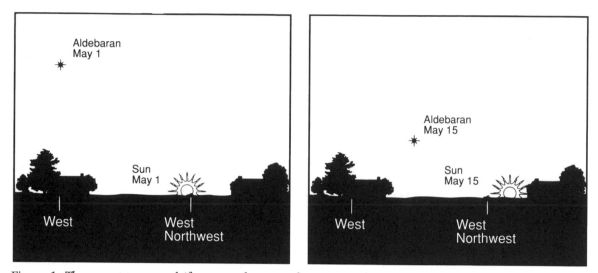

Figure 1. *The sun appears to drift eastward among the stars, making one complete circuit through the sky each year. Throughout May, for example, the setting sun creeps ever closer to the bright star Aldebaran.*

Passover are similarly tied to the occurrence of the spring equinox, a date when the hours of daylight equal the hours of darkness. In a more secular vein, Groundhog Day is a "cross-quarter day," roughly the midway point between winter solstice and spring equinox. Likewise Halloween falls in the midst of the dark time between the autumnal equinox and the winter solstice, when daylight reaches its minimum.

The moon, a lesser light, may be what first draws our attention to the night sky. Naturally, the ancients held it in high regard. Early calendars were based on the motions of the moon, mainly because its cycle is shorter and more obvious than the sun's. The priest-astronomers of ancient Mesopotamia, for instance, determined the start of a new lunar month by watching for the first appearance of the moon's slender crescent in the evening twilight.

But even before observations of the sun and moon became part of a formalized time-keeping system, ancient observers must have realized that the sun and moon weren't the only objects moving through the heavens. Unlike most moderns, the diligent star-watchers of ancient times also noticed five peculiar "stars" that wander independently among the constellations, each moving at different speeds. Like the sun and moon, they follow roughly the same path through the sky and usually move eastward relative to the stars. But now and then these wanderers slow down, stop, loop west for a time, stop again, and then resume their eastward journey. Today we know these objects as the planets, a word which derives from the literal Greek description of their peculiar astronomical behavior (*planetes*, "wanderers"). Ancient astronomers knew of only five planets: Mercury, Venus, Mars, Jupiter, and Saturn. Although under ideal conditions Uranus can be seen without some sort of optical aid, it evidently didn't catch the attention of early sky-watchers. Uranus, like still fainter Neptune and Pluto, would remain undiscovered

until long after the invention of the telescope.

While there's no question that the sky played an important role in the great cultures of the past, we take for granted the vestiges of its significance remaining in today's society. The week has seven days, each of which is named for one of the seven objects known to move through the sky since ancient times. Our English nomenclature dates back to Roman names for the weekdays. Sunday derives from *Dies Solis*, "Day of the Sun," Monday comes from *Dies Lunae*, "Day of the Moon," and Saturday is Saturn's Day. The remaining celestial connections are filtered through the pantheons and languages of other peoples: Tuesday (Mars), Wednesday (Mercury), Thursday (Jupiter), and Friday (Venus).

Watching the Sky

Now it's your turn to consider the heavens from a modern perspective. First, let's examine the phenomena — what do we *see*? The most obvious observation is that of the daily motion of the sun, moon, and stars from east to west. This simple motion is, itself, explainable two ways: either the sky is spinning around us, or the earth itself is turning. Extended, methodical observation reveals several other motions superimposed on this basic east-to-west movement.

One type of motion is best observed around sunset (Fig. 1). As the twilight fades, note the position of the first star you see to the west or south. Observe that star at the same time for a few days, and things seem quite in order. But after a week or two you'll notice that the star drifts slowly westward from its original position. Over the course of several weeks you'll find that it appears ever closer to the sun until it's eventually lost in the glow of twilight.

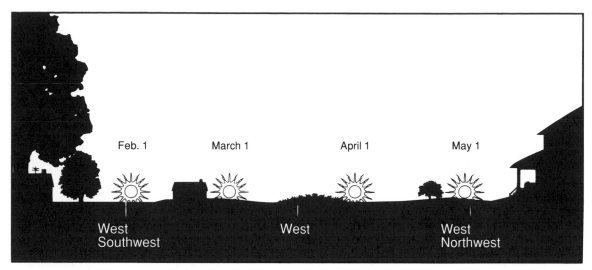

Figure 2. Track the setting sun for a few weeks and you'll see the direction of sunset move along the horizon. The sun slides northward from late December to late June, then reverses direction and heads south for the rest of the year. This diagram shows how the sunset point travels between February and May for observers in middle northern latitudes.

Of course, this westward drift of the stars can be viewed the other way around — as an eastward drift of the sun through the stars. Ancient observers noticed that the sun made a complete circuit through the stars once a year. The path it traveled became known as the *ecliptic*. The twelve constellations along the sun's path are the famous astrological "signs of the zodiac," yet another vestige of our prescientific ways.

Consider yet another motion of the sun, noticeable right at sunset (Fig. 2). Look once a week, marking the place on the horizon where the sun disappears. After a few weeks you'll find that the direction of sunset seems to drift along the horizon: southward from July to December, northward the other half of the year. The farthest positions north and south correspond to the summer and winter solstices, respectively, and the point midway between them marks the sun's position at the spring and autumnal equinoxes.

These motions — the sun through the stars and the sunset along the horizon — are in fact reflections of the earth's motion around the sun. In the first case, earth's motion merely places the sun closer to our line of sight with the star (Fig. 3). Eventually, both the sun and star lie roughly along the same line of sight and the star is lost in the sun's brilliant glare. When the star emerges from twilight on the opposite side of the sun, it will be visible just before sunrise.

Figure 4 explains the north-south drift of the sunset point. Earth spins at an angle to its path around the sun, and this tilt affects our point of view throughout the year. When the northern hemisphere is fully angled into the sun on the first day of northern summer, the sun rises and sets

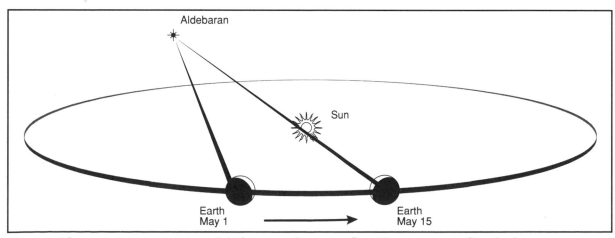

Figure 3. The sun's apparent eastward drift among the stars reflects the movement of earth along its orbit. Once each year, for instance, earth's motion in space brings the sun close to our line of sight with distant Aldebaran.

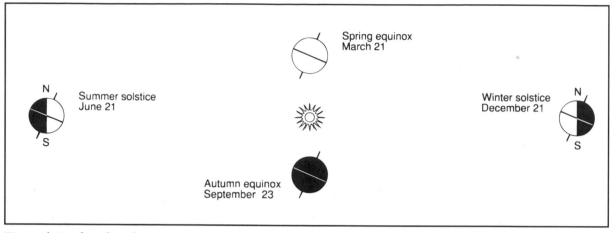

Figure 4. *Earth's tilt with respect to its orbit creates the seasons. On June 21, the start of northern summer, the northern hemisphere angles into the sun most directly. The sun arcs highest in the sky, rising and setting at its northernmost points. Six months later, on December 21, northern observers see the sun make its lowest track through the sky, rising and setting at its southernmost points.*

at its northernmost point and reaches its highest altitude at noon. Six months later, on the first day of northern winter, earth's southern hemisphere faces into the sun. The sun, now rising and setting at its southernmost point, makes its lowest arc across the northern sky and creates the long shadows of winter.

The moon exhibits a similar dual motion — east to west throughout the night, but west to east through the stars. This latter motion is much more obvious than the sun's, for the moon courses through the entire celestial vault in less than a month. We'll discuss these motions more fully in Chapter 1.

The stars are the backdrop of all we'll observe in this book. They form a sort of road map of the heavens, the signposts by which we mark the motions of the bodies of the solar system. They arc across the sky as part of their daily motion east to west and, as Figure 1 shows, drift toward evening twilight for an annual rendezvous with the sun. Unlike the sun and moon, a given star rises or sets at the same location on the horizon day after day, season after season, year after year. And, for practical purposes, the positions of the stars are fixed in relation to each other. We'll turn again to the stars in Chapter 6.

Besides the relative motions of the denizens of the sky, there's one more outstanding characteristic of celestial objects: their brightness. One of the first truly scientific acts of the ancients was to make rigorous charts of the heavens. The astronomer most notable for such careful observations was the Greek Hipparchus. Chief among his contributions is the system by which we rank the brightness of celestial objects. In 129 B.C. Hipparchus cataloged the stars visible from his home, noting their position and brightness. His notation was a model of clarity: The brightest stars were of the "first magnitude," the next brightest were second magnitude, and so on, down to sixth magnitude, the faintest visible to the unaided eye. Although described this way the brightness scheme makes sense, it often confuses new observers that a star with a higher magnitude number is actually fainter than one with a lower value. That is, a magnitude 1 star is brighter than a magnitude 2 star. If this gives you trouble, think of the word "magnitude" as meaning "class."

When the system was later precisely codified, it became clear that some of Hipparchus's first magnitude stars were considerably brighter than others. So astronomers found it necessary to extend the magnitude scale to zero and beyond — Sirius, the brightest star, has a magnitude of −1.4! When we apply this scale to objects in the solar system, it gets even more peculiar: Mars can attain a brightness of −2.8; Venus at its brightest is magnitude −4.6; the full moon ranks as −12.6; and the sun tops the scale at −26.8! The astronomical brightness scale is a bit counterintuitive and we'll discuss it in more detail in Chapter 6. For now, just keep in mind that the smaller the magnitude value — including those "below zero" — the brighter the object seems.

We've just given you a whirlwind introduction to the motions of the heavens. We encourage you to spend some evenings examining these motions for yourself. As you do, you'll get a clearer view of what we've described. And of course, you're sure to notice things we haven't mentioned.

That's what the rest of the book is about!

1

MOON DANCE

The moon is our nearest neighbor in space, earth's partner in a never-ending dance around the sun. For much of each month, it's also the most prominent object in the nighttime sky. The moon whirls around the celestial vault in just over twenty-seven days, and each day throughout its cycle appears slightly different — a sliver of light at the beginning and end, a brilliant disk in the middle. Compared to every other object in the sky, the moon is a vagrant show-off. Its regular cycle of waxing and waning is so obvious that it formed the basis of early calendars, providing a convenient interval of time intermediate between the day and the more subtle annual cycle of the sun and seasons. While there is only one purely lunar calendar in wide use today, the moon remains an important cultural symbol. The flags of more than half a dozen nations feature its crescent, and some of the world's most important religious festivals are tied to the phases of the moon. As both a signpost of the cosmic and a symbol of inaccessibility, the moon has been a source of inspiration and mystery throughout the ages.

Today, however, the moon reminds us of a remarkable technological milestone. A dozen people have walked on its dusty gray surface, placing scientific instruments, retrieving rock samples, taking photographs — and in one case even hitting a golf ball! The moon is the first and only alien world humanity has visited in person. It's fitting, then, that this stepping-stone of the space age should also serve as the starting point for your personal explorations of the solar system.

By the Light of the Moon

The moon dazzles us, illuminating the night and washing out the faintest stars, yet it ranks among the darkest bodies in the solar system. Like the planets, the moon shines by reflecting the light of the sun — the more light it mirrors,

the brighter it appears. The moon reflects just 7 percent of the sunlight it receives, which makes its surface about as reflective as asphalt! Our impression of the moon as a brilliant object stems in part from the contrast between its sunlit disk and the dark night sky, and in part from the sensitivity of the human eye once it adapts to darkness. The dim daytime moon, barely visible against the bright blue sky, gives a far more representative picture of our satellite's intrinsic brightness.

Another of the moon's more distinctive qualities is its rapid motion through the sky. It completes one orbit around the earth every 27.32 days on average. This period, called the *sidereal month*, represents the time it takes the moon to make one full pass through the sky with respect to the stars. If on some night we find the moon in the vicinity of a bright star, a sidereal month must pass before the moon returns to that area of the sky. The moon's orbital motion carries it from west to east by about thirteen degrees per day — an angle greater than the apparent width of your fist held at arm's length. On occasions when the moon lies near a bright star or planet, its eastward motion may be apparent even in the course of an hour (Fig. 1-1, cover photo). As a result of this rapid eastward motion, the moon rises an average of fifty minutes later each day. The earth must spin us about fifty minutes longer every day before we're turned toward the moon's new position.

Such phenomena are admittedly a bit too subtle for most of us, but the moon's ramblings also create a far more conspicuous display. Even the most casual skywatchers don't fail to notice that the moon goes through a sequence of changing shapes, or phases. Because of the demands of our lifestyles, we usually notice the moon during the first half of this cycle, beginning with its thin crescent in the western twilight and growing to the full moon rising in the east at dusk. Half of the moon is

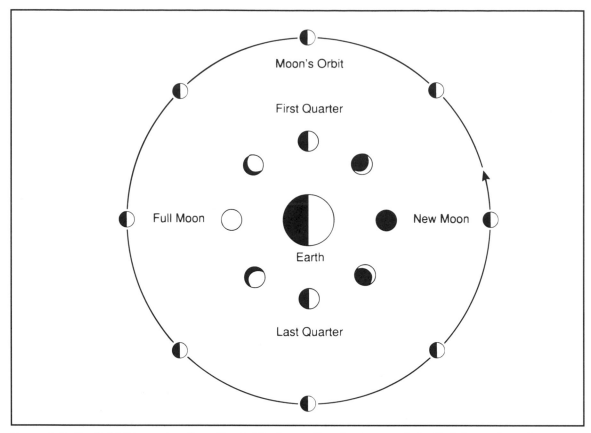

Figure 1-2. *Sunlight (streaming from the right) always illuminates exactly half of the earth and moon. The moon's motion along its orbit (outer circle) allows us to see different fractions of its sunlit hemisphere, resulting in the familiar lunar phases (inner circle).*

always in sunlight, but we see it go through phases because the portion of the sunlit hemisphere that faces earth changes during the course of its orbit. (Fig. 1-2).

Astronomers refer to each complete cycle of phases as a *lunation*. The cycle begins at *new moon*, when the moon is positioned roughly between earth and the sun. Since the side facing away from earth receives all of the sunlight, the new moon cannot be seen unless it passes directly in front of the sun (a solar eclipse). Usually within about twenty-five hours (but sometimes less than fifteen), the moon's motion carries it far enough east of the sun that attentive observers can find its hair-thin crescent low in the west just after sunset.

As the moon moves further east, it rises and sets progressively later than the sun and also shows us a greater portion of its daylight side. After about a week that slim crescent has grown to a half-moon — the phase somewhat confusingly known as the *first-quarter moon* (because it occurs one-quarter of the way through the monthly cycle). It rises around noon, sets near midnight, and is highest in the sky at sunset. In another week the earth, moon, and sun again align — but this time the earth lies roughly in the middle. The *full moon*, now opposite the sun in the sky, rises in the east as the sun sets in the west and is visible all night long.

Up to this point the moon is said to be waxing, or

growing, but after full moon it begins to wane. Within about a week it shrinks to the half-disk of *last-quarter moon*, which rises around midnight, stands highest in the sky at dawn, and sets at noon. Throughout the last week of a lunation the moon becomes a thinner and thinner crescent and rises ever closer to the sun. The cycle begins anew when earth, moon, and sun realign for new moon. (See the Appendix for the times and dates of lunar phases.)

We noted above that the moon takes 27.32 days to circle the earth, but each lunation cycle takes an average of 29.53 days. Why the discrepancy? The moon completes an orbit in one sidereal month, but during that time the earth and moon together have moved about one-twelfth of the way around their yearly orbit of the sun. Relative to the background stars, the sun appears to have moved about twenty-seven degrees to the east of where it was at the start of each lunation. Before new moon can again occur, the moon must travel a bit farther in its orbit to line up with the earth and sun. Since it takes the moon a couple of days to make up the difference, the lunation cycle takes that much more time. This longer period, called the *synodic month*, forms the basis of all lunar calendars.

Take another look at Fig. 1-2. In this diagram, the sun lies to the right and the earth's orbital motion carries it downward. When you gaze at the moon during either of

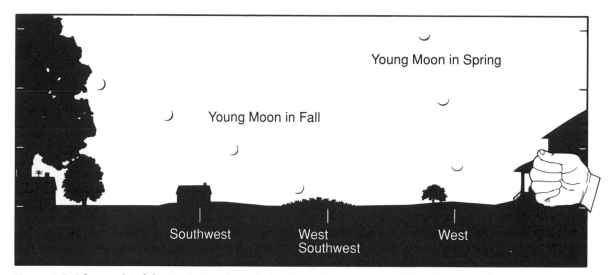

Figure 1-3. *The angle of the moon's path to the horizon changes throughout the year. The young moon can be seen more easily in the spring evenings than in the fall; the reverse is true for the waning crescent seen in the mornings. For clarity, the moon is enlarged four times. Horizon scenes like these occur throughout the book. Tick marks along the bottom mark off the directions, and those along the sides represent altitude intervals of ten degrees — equal to the width of your fist at arm's length.*

its quarter phases, you're looking right along our planet's direction of motion around the sun. The first-quarter moon marks the approximate position occupied by the earth about three-and-a-half hours earlier; the last-quarter moon shows you earth's location at roughly the same time in the future.

The moon's crescent phases offer another observing treat. For a few days at the start and end of each lunation, we can see more of the moon than its thin, sunlit crescent — the remainder of the disk also glows faintly. Sometimes called "the old moon in the new moon's arms" when seen in the evening, this secondary illumination is caused by sunlight reflected from the earth. From the moon our planet is a brilliant bluish disk almost four times larger than the full moon we see and, thanks to clouds, ice caps, and oceans, it reflects more than five times as much sunlight. In addition, the phases of earth and moon are complementary — when we see the crescent moon, a lunar astronomer would see an almost-full earth. Leonardo da Vinci, the fifteenth-century Italian artist and engineer, seems to have been the first to explain the phenomenon correctly. "Earthshine" is most noticeable on the young moon during spring evenings, when the ecliptic cuts the horizon most steeply and the setting moon stands high. For the same reason, autumn mornings are best for seeing earthshine on the old moon (Fig. 1-3).

A more curious phenomenon can be seen around the time of full moon. As it rises above the buildings and trees that line the local horizon, the moon's disk seems much larger than it does when it appears higher in the sky a few hours later. Yet actual measures of the moon's angular diameter at these times show that it remains the same — one-half of one degree. The "moon illusion"

occurs because we unconsciously compare the rising moon to nearby references — a house, a car, a tree — that are not so compelling when the moon reaches higher altitudes.

Time and Tide

The earliest complete calendars were developed by 2000 B.C. and based on the phases of the moon. Days began at sunset, and religious authorities declared a month to have begun when they first saw the young crescent low in the west; if poor weather obscured the moon, then the start of the month was determined by calculation. Since the calendar was based on the phases of the moon (its synodic period), each month held either twenty-nine or thirty days, alternating — the basis of the month in today's Western calendar. The moon served as a convenient timepiece, but there are built-in problems with any purely lunar calendar. Twelve complete cycles of the moon typically take 354 days, eleven short of the seasonal year. Left to itself this mismatch would cause each year to begin about eleven days earlier, slowly drifting back through the seasons. In fact, this is exactly the case with the traditional Muslim calendar, the only true lunar calendar still in wide use. The beginning of the year slides through the seasons for about thirty-three years, after which things are back where they started.

Most cultures found this discrepancy much harder to accept because the seasonal year controlled agricultural activities. Early astronomers spent considerable effort devising methods of intercalation — the addition of "leap" days or months — to make the lunar calendar jibe with the solar year. A familiar example is the 366th day (February 29) added every four years to the Western calendar so that the average length of each year more closely matches the earth's orbital period.

Such corrections were probably first made simply by watching the harvest and adding whatever length of time seemed necessary; later on, other celestial cycles were recognized that facilitated the process. One such cycle, also involving lunar phases, is named for the Greek astronomer Meton of Athens. In 433 B.C. Meton noticed that 235 synodic months occurred in almost exactly nineteen seasonal years, which meant that the moon's phases recur on essentially the same dates every nineteen years. For example: Full moons occur on January 8 in 1917, 1936, 1955, 1974, 1993, 2031, and 2050, and in 2012 on January 9. The earth's annual motion around the sun gives rise to the seasons, and the monthly relationship between the sun and moon causes phases — and since the Metonic cycle relates one to the other, it could be used to keep a lunar calendar in synch with the seasons.

Vestiges of the moon's importance as a timekeeper survive in the timing of religious festivals. The Koran specifically instructs Muslims to see the young moon before beginning the daylight fast of Ramadan. Rosh Hashanah, the first day of the Hebrew calendar, falls on the new moon after the September equinox, although certain calendar rules may cause it to be postponed for a couple of days. And the Christian festival of Easter falls on the first Sunday following the first full moon after the March equinox — so it can occur as early as March 22 and as late as April 25.

Of course, not everyone based their month on the phases of the moon. In India an early calendar gave the moon's motion through the stars (that is, the sidereal period) more importance, so the months contained twenty-seven or twenty-eight days. The Inca civilization of South America also worked the sidereal month into their calendar. Farther north, the Maya perfected a uniquely nonlunar calendar.

Nearly everywhere, however, the moon was correctly linked to a primary natural rhythm — the ebb and flow of the tides. "The moon draws the sea after it with a powerful suction," wrote Pliny the Elder, a Roman naturalist, in the first century A.D. This association arose because of the compelling coincidence between the moon's motion and the cycle of tides: moonrise occurs every 24 hours and 50 minutes, on average, and the period between successive high tides averages 12 hours and 25 minutes. Galileo Galilei and Johannes Kepler, two of the most prominent scientists of the seventeenth century, simply didn't buy the idea, but the proof came not long after them. In 1687 the English physicist Isaac Newton published what many consider his single greatest work, a book entitled *Philosophiae naturalis principia mathematica* (*Mathematical Principles of Natural Philosophy*), in which he describes the theory of gravity that allowed him to "deduce the motions of the planets, the comets, the moon, and the sea."

Just as the gravitational attraction of the sun prevents the planets from flying off into space, so the gravity of the earth keeps the moon in its orbit. But the moon pulls back on the earth, creating a tidal swelling on opposite sides. Think of these tidal bulges as the crests of a giant wave, low in height — only a few feet in open ocean — but of tremendous length. The crests of these two waves move over the earth's surface as the moon proceeds in its monthly orbit. The earth spins beneath them so that a given location passes under each tidal bulge once every 24 hours and 50 minutes (the time between successive risings of the moon). Since there are two bulges, a high tide occurs every 12 hours and 25 minutes. This simple picture is greatly complicated by the positions of continents, the irregular shapes of ocean basins, and friction.

The sun also raises a pair of tidal bulges and, while they have less than half the height of those raised by the moon, there are times each month when solar and lunar tides reinforce and interfere with one another. When the earth, sun, and moon are aligned during full and new moon, a higher-than-normal high tide (called a spring tide) occurs because the solar and lunar tidal crests coincide. At the moon's quarter phases, solar tides instead interfere and a lower-than-normal high tide (neap tide) results. These extreme tides may change the water level by up to 20 percent above or below normal tidal limits.

Naturally, the earth's gravity also affects the moon. Thanks to a slight bulge of matter on the moon's near side, over time the earth's pull altered the moon's rotation until its spin became synchronous, or "locked," to its orbital period. This means that the moon spins once on its rotational axis every orbit, or completes one rotation every sidereal month. As a consequence, the moon always shows us the same face as it orbits the earth. When people speak of the "dark side" of the moon they're really referring to its far side, the side we never see from earth; every part of the moon is illuminated for half the lunar month. (For example: The far side of the moon is in daylight at new moon, but the side facing earth is dark.)

If the moon has influence over the vastness of the sea, might it also have power over life on earth? This idea, central to many occult themes, is reflected in folk tales of mythical creatures like werewolves. But many animals that live in or near the ocean *do* exhibit cycles of activity patterned by the rhythm of the tides and the phases of the moon. During the spring and early summer, for example, a small coastal fish called the grunion spawns by the thousands on the beaches of southern California a few days after full and new moon. Samoan palolo worms swarm during the moon's last quarter in October and November. Under laboratory conditions, with no cues from water movements or light from the sun or moon, fiddler crabs show a decidedly lunar activity cycle of 24 hours and 50 minutes.

Is there a "biological tide" in humans? The moon is commonly believed to influence human behavior, an age-

Figure 1-4. *The moon's largest surface features are easily recognized with the naked eye and play a role in lunar folk tales. The popular Man in the Moon (left) is most obvious early in the evening, when the rising full moon appears tilted to the left. The markings resemble the profile of an old woman (center) when the moon is highest in the sky, and the Rabbit (right) adorns the setting moon.*

old idea reflected in such words as *moonstruck, moon-eyed, moony,* and *lunatic.* Many people having an unusually eventful day will chalk it up to the full moon, but few will bother to see what the lunar phase really is. The moon's long association with fertility stems from a tantalizing coincidence between the average length of the human menstrual cycle and the moon's sidereal month. Could there be some lunar effect on the brain? A study of homicide statistics in the 1970s seems to show an increase in murders around the new and full phases, but subsequent larger studies failed to find any link between the moon's motion and crime, suicides, or psychiatric admissions.

Moon Lore, Moon Science

The moon has been worshiped and personified in many different ways, although it was not usually connected with a powerful deity. Various stories arose to explain its movements and appearance. We often picture the moon's dark markings as forming the Man in the Moon. According to one story, he was caught working on Sunday and cast onto the moon — perhaps that accounts for his expression of surprise. Other commonly recognized figures include a rabbit, a frequent symbol for the moon in ancient America, and the profile of a girl or old woman (Fig. 1-4). One Scandinavian story about a boy and girl swept up by the moon appears to have been the basis of the nursery rhyme of Jack and Jill.

The Babylonians saw the moon as *Sin,* a wise god who marked off time and whose advice was sought by other deities each month. His bright light kept evil forces at bay during the night, but he was attacked by these creatures when they conspired with Sin's daughter *Ishtar* (the planet Venus) to extinguish his light; *Marduk,* the chief Babylonian deity, fought them off and restored Sin. Another ancient moon god, the Egyptian *Thoth,* was the

inventor of mathematics and hieroglyphics, and one of those who weighed the hearts of the dead as they were tried in the afterlife; the gods *Khons* (male) and *Isis* (female), also associated with the moon, were reputed to have special healing powers.

In Greece, *Artemis* ruled the moon, the hunt, and all nature, brought fair weather for travelers, and protected young girls. She later became identified with *Selene,* who fell in love with the shepherd Endymion and visited him nightly; her fifty daughters represent the fifty synodic months between the Olympic Games. The Roman moon goddess *Diana,* identified with Artemis, became the patroness of witches in medieval Europe.

By the fifth century B.C., Greek astronomy and mathematics had become sophisticated enough for some imaginative scientists to explore the moon's true nature. Anaxagoras (ca. 500–428 B.C.) correctly explained the causes of eclipses and was banished for considering that the moon was at least partly made of the same material as the earth. Both he and the influential philosopher Aristotle (384–322 B.C.) said the moon was a solid sphere illuminated by sunlight and explained its phases. Using simple geometry, Aristarchos (ca. 320–250 B.C.) showed that the moon was much closer than the sun, and Hipparchus (ca. 190–125 B.C.) later determined the moon's distance within 10 percent of the correct value. Despite these pioneering efforts, the moon came to be regarded as a sphere of crystalline smoothness and perfection, a member of the flawless realm beyond the mundane one.

That all changed in 1609, when Galileo Galilei became the first to explain the moon through a new invention — the spyglass, later named the telescope. Although only as powerful as a good pair of modern binoculars, Galileo's spyglass revealed features resembling jagged peaks, valleys, and pockmarked plains. He studied these features through the moon's changing phases and recog-

Table 1-1. Facts About the Earth and Moon

Earth

Diameter:	12,756 kilometers 7,927 miles
Average surface temperature:	15° C 59° F
Atmospheric surface pressure:	1 bar
Atmospheric composition:	77% nitrogen 21% oxygen 1% water vapor 0.9% argon
Rotation period:	23.93 hours
Sidereal orbital period:	365.26 days or 1 year
Mean distance from sun:	149.60 million kilometers 92.96 million miles

Moon

Diameter:	3,476 kilometers 2,160 miles 27% that of earth
Surface temperature:	-175° to 130° C -283° to 266° F
Sidereal orbital period: (true period of rotation and revolution)	27.32 days
Synodic orbital period: (time between repeating phases, e.g., new moon to new moon)	29.53 days
Mean distance from earth:	384,400 kilometers 238,870 miles
Orbit inclined to earth's:	5.14°

nized their three-dimensional character by the shadows they cast. This confirmed his belief that "like the face of earth itself [the moon] is everywhere filled with protuberances, deep chasms, and sinuosities." The plains coincided with the moon's dark markings and Galileo called them *maria*, the Latin for seas (the singular is *mare*, pronounced MAH-ray). Naturally enough, he termed the rest of the Moon's surface *terrae* (lands), although today they're more commonly called the lunar highlands.

Galileo's observations were an important step in proving the Copernican view that the earth was an ordinary member of the solar system. Isaac Newton, similarly inspired by the moon, extended this line of reasoning to conclude that the force that bought apples to the ground was the same that kept the moon circling the earth and the planets around the sun. As astronomers discovered satellites orbiting other planets, they began to appreciate the uniqueness of our own — the largest in the solar system relative to the planet it orbits. Slightly more then one-fourth the diameter of the earth, the moon has a total surface area sightly larger than Africa and contains less than one-eighteenth the earth's mass. This makes the force of gravity on the lunar surface one-sixth that of earth's. This is too weak for the moon to hold onto gases escaping from its interior, so the moon has no atmosphere. Table 1-1 lists some basic facts about the earth and moon.

After Galileo announced his telescopic discoveries, cartographers set to work mapping lunar features. One of the first maps of the entire visible hemisphere was published in 1645 by Langrenus, an astronomer in the court of King Philip IV of Spain. He identified about three hundred features (mostly craters) and showed a certain political savvy by naming them for assorted kings and noblemen, for example, Louis XIV and Philip IV. A more palatable but equally arbitrary nomenclature arrived with the lunar map of Giovanni Riccioli in 1651. He gave the maria Latin names reflecting qualities or characteristics (Sea of Tranquility, Sea of Serenity, Sea of Cold, Sea of Nectar) and early ideas that connected the moon with weather changes (Sea of Rains, Sea of Vapors, Sea of Clouds, Ocean of Storms). Craters were named for scholars and scientists (Copernicus, Tycho), and mountain ranges — really the rims of vast, ancient craters — were named after famous terrestrial ranges, such as the Alps or Apennines. Figure 1-5 locates some of the moon's most prominent features.

We've come to understand much about the moon's complex history and the violent origins of the solar system. From 1969 to 1972, six American Apollo missions brought a dozen men to the moon's surface and 843 pounds (382 kilograms) of lunar soil and rocks to the earth's (Fig. 1-6). Between 1970 and 1977 Soviet robot probes provided another 11 ounces (300 grams) of surface material from three additional locations. The airless, waterless surface of the moon has preserved many clues to the solar system's past.

Seen through a telescope or from orbiting spacecraft, the lunar highlands break up into an endless series of overlapping meteorite craters. These regions took the brunt of a bombardment that formed the solar system's moons and planets through powerful collisions. The top few kilometers of the moon's surface has been repeatedly mixed and pulverized. The lunar highlands, the moon's most ancient terrain, contain samples that solidified within an ocean of molten rock about 4.4 billion years ago — about 500 million years older than the oldest rocks found on earth.

The dark lunar "seas" represent more recent terrain. The maria cover about sixteen percent of the moon's surface, mostly on the hemisphere that faces earth. As the moon cooled and its crust solidified, several giant impacts formed huge multi-ringed basins between 4.0 and 3.8 billion years ago; they were later flooded with lavas oozing from the moon's interior. The maria are the frozen remains of ancient lava flows that erased the older craters beneath them, which explains why they have fewer craters than the neighboring highlands. With the exception of a few large craters, such as Tycho (54 miles or 87 kilometers across) and Copernicus (57 miles or 91 kilometers), and many smaller ones, the moon's face has changed little in the past billion years. According to geologists, the only activity now left for the moon is an occasional shudder from deep moonquakes or impacting meteorites; it is all but dead geologically.

There is one lunar phenomenon for which no has an adequate explanation. Puzzling glows, hazes, color changes, and brief obscurations of features have been reported around certain places on the moon for centuries. Given the great age of the lunar lavas we see, it's difficult to imagine that these so-called transient lunar events could be true volcanic eruptions. They may be caused by small amounts of gas escaping through cracks in the moon's surface.

As astronomers became more and more familiar with the moon, a nagging problem remained: Where did it come from? Prior to the Apollo missions, planetary scientists advanced several theories on the origin of the moon. One idea was that the moon was born simultaneously from the same cloud of solar system debris as the earth. That theory, however, required that the moon be a miniature version of the earth, made of the same ratio of rock and metal; but we now know that the moon contains relatively little iron and nickel. Another idea proposed that the moon was a wandering spacerock captured by earth's gravity — but the probability of such an event is so vanishingly small that most astronomers found it untenable. Yet another theory was that the earth once spun so rapidly that the moon was simply thrown off. This "fission theory" requires that the earth-moon system today hold more angular momentum than we actually observe. Naturally, it was hoped that analysis of actual lunar rocks would eventually favor some of these ideas and rule out others. In fact, no pre-Apollo theory of the moon's birth adequately fits our new knowledge of its chemical and geological make-up.

Lunar rock samples have given scientists a definite set of features to explain. The moon appears to be made of rock that is basically similar to earth's but lacks compounds that normally accompany iron and compounds that vaporize at low temperatures. During the past fifteen years or so, astronomers have converged on the idea that a planet-sized piece of debris roughly the size of Mars made a grazing collision with Earth shortly after its creation. Such a collision would blast rock from the surface, heat it intensely, and form a disk of material around Earth. Within a few tens of millions of years such a disk of debris would coalesce into an object that matches lunar mass, density, and composition.

Long an inspiration to poets and lovers, the moon remains a symbol of mystery and an eerie beacon of

Figure 1-5. *Above: The full moon. Dark regions are the lunar "seas." Facing page (top): Features labeled on this map may be seen easily with binoculars. The craters Tycho and Copernicus are two of the moon's newest; light-colored debris from these impacts form rays that spread far from the craters. (Photo by Bill Sterne)*

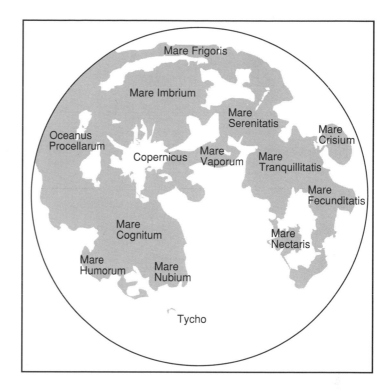

other-worldliness. It is also an invitation to explore the sky, to ponder the ways of the universe, and to inquire about the realms beyond our familiar planet. Best of all, for the beginning stargazer, it is on display year-round. As you read on and search out the other members of the solar system's family, remember to enjoy the rhythms of our "next-door neighbor."

Figure 1-6. *Apollo 17 astronaut Eugene Cernan takes a spin on the lunar surface in December, 1972. (Photo by NASA)*

2 MORNING STARS, EVENING STARS: VENUS AND MERCURY

Like the ancient astronomers, we may be captivated by the lovely sight of the crescent moon on the western horizon at dusk. If we followed its movements faithfully it would lead us, sooner or later, to other discoveries. Some morning just before new moon, or some evening just after it, the slim crescent will share the twilight with a star that outshines all others. If the timing is just right, a second speck of light — fainter, a bit redder — will waver in the unsteady air close to the horizon. With that observation your grasp of the universe swells over two hundredfold, reaching far beyond the orbit of the moon to encompass the planets Venus and Mercury.

These two "stars" never appear more than a few hours behind or ahead of the sun. They emerge into the evening or morning twilight only to reverse course and, like celestial moths, flit back toward the light of the sun. The familiar terms "morning star" and "evening star" hardly carry with them the weight of scientific precision. They have come to mean a planet bright enough to stand out in the glow of dusk or dawn. All of the classical planets will sooner or later satisfy this condition, but only two spend most of their time in the morning and evening twilight. So the titles "evening star" and "morning star" are really best conferred upon Mercury and Venus, the planets whose movements through the sky are most closely tied to the sun.

Actually, Venus so outshines every other planet that it deserves a title all its own. It's the third brightest object in the sky and can be seen even during the day under ideal observing conditions. At its best Venus swings far enough from the sun to leave the twilight behind; for weeks on end skywatchers will see it as a gleaming jewel set high in a darkened sky. The orbit of Venus also brings it within about twenty-five million miles of Earth, or only about one

hundred times the distance to the moon. No other planet ever passes as close to Earth. All this adds up to one simple fact: Anyone can find Venus knowing only the most basic information about when and where to look for it.

None of that, however, can be said of Mercury. Although it mimics the apparent sky motions of Venus, Mercury lacks both its sweep and brilliance. Even at its best Mercury hugs the horizon, shining weakly through the haze and twilight. Moreover, Mercury's best appearances don't last very long. It bobs up above the predawn horizon for a week or so, retreats back toward the sun, and then shows itself briefly in the west after sunset. To make matters even worse, the time of year also critically affects the planet's visibility. Of course, we prefer that you look at the situation this way: Mercury presents an observational challenge unmatched by any of the other planets visible to the unaided eye.

The charts and tables in this chapter will help you find both Mercury and Venus. Sometimes the two planets shine together in the morning and evening twilight; Venus then acts as a celestial landmark, pointing the way to the more elusive Mercury. The crescent moon and other planets may join them too, creating arrangements that make interesting astronomical "photo opportunities" (Fig. 2-1).

Since Venus is the brightest, closest, and most visible of all the planets, it's the logical place to start. Venus has fascinated stargazers and scientists alike for centuries — but before we discuss that, why not observe this planet for yourself! Turn to Table 2-1, which summarizes the visibility of Venus as morning and evening star. If it's not observable on the date you read this, make plans to look for it at the next available opportunity. Seeing is believing!

Figure 2-1. *The crescent moon hangs near Venus and Jupiter on October 4, 1991. Venus, the brightest "star" in this picture, passes closer to Earth than any other planet and ranks as the third brightest object in the sky. Compare this arrangement to some of the finder charts in this chapter. (Photo by Francis Reddy)*

Queen of the Heavens

Venus was probably the first planet noticed by ancient cultures. The astronomers of Imperial China knew it as *T'ai po*, the "Great White One," while in India it was known as *Sukra*. Our name for the planet comes from the Romans, who identified it with their goddess of love and beauty, but it's the Norse fertility goddess *Freya* who lends her name to the day ruled by Venus. "Freya's day" is Friday.

Venus was an obscure deity in ancient Rome until she became associated with the powerful Greek goddess *Aphrodite*. Julius Caesar himself enshrined her as *Venus Genetrix*, the ancestor of his own family. Because Venus alternates between evening and morning appearances without remaining visible all night long, some cultures knew the planet by one name when it appeared in the west and by another when it shone in the east. Possibly some thought they were seeing two different objects. The early Greeks called the bright evening star *Hesperos* and the morning star *Phosphoros*. The Roman equivalents were *Vesper* and *Lucifer*.

The practice of associating the "wanderers" with the most important gods originated in Babylonia. There, Venus was considered a manifestation of *Ishtar*, a powerful fertility goddess, and known by the earlier Akkadian names *Dil.bad* or *Nin.si-an.na*. Records dating back to

the First Dynasty — nearly four thousand years ago — show that Babylonian observers were thoroughly familiar with the details of the planet's visibility cycle. The evidence comes from a list of dates for the beginning and end of each of the planet's morning and evening appearances during the twenty-one-year reign of King Ammisaduga. Appended to the observations are astrological predictions such as "a king will send messages of peace to another king" or "the heart of the land will be happy."

Two millennia later, the Mayan civilization in what is now Belize, Guatemala, and southeastern Mexico independently discerned the pattern of Venus's motions. They knew Venus by several names — Great Star, Wasp Star, Bright Star — and associated it with *Kukulcan* ("feathered serpent"), god of the wind and inventor of the calendar. Its movements were of such astrological importance that the Maya oriented at least two major buildings with Venus in mind — the Caracol at Chichen Itza and the Governor's Palace at Uxmal, both on the plains of Mexico's Yucatan Peninsula. They devote six pages to Venus in the Dresden Codex, one of a handful of surviving documents, and include detailed astronomical and religious information. They greatly feared the planet in the days just after its switch to the morning sky. Illustrations in the Codex show Venus deities flinging spears at various earthly victims, with singularly unpleasant prognostications nearby — "Woe to the maize, woe to the corn fields, drought,

Table 2-1 Visibility of Venus, 1993–2001. Each entry gives the range of dates for which an observer at mid-northern latitudes will find Venus at least ten degrees above the horizon forty-five minutes before sunrise or after sunset.

Year	As Evening Star (In the west after sunset)	As Morning Star (In the east before sunrise)
1993	Early November 1992 to mid-March. Best in early February, when it sets nearly 4 hours after the sun.	Early May to late October. Best in early August, rising about 3 hours before the sun.
1994	Early April to late August. Best in early June, but it remains low in the northwest and sets just 2½ hours after sundown.	Mid-November to mid-March 1995. Best in late December, rising about 3¾ hours before dawn.
1996	Mid-December 1995 to late May. Best in late March, setting 4 hours after sundown.	Early July to late December. Best in early September, rising about 3¾ hours before dawn.
1997	Mid-October to early January 1998. Best in early December, Venus sets 3 hours after the sun.	
1998		Early February to early May. Venus rises a little more than 2¼ hours before dawn and remains low in the southeast.
1999	Late January to mid-July. Best in mid-May, setting about 3½ hours after the sun.	Early September to late January 2000. Best in late October, rising nearly 4 hours ahead of the sun.
2001	Early November 2000 to mid-March. Best in early February, when it sets nearly 4 hours after the sun. This is a repeat of the 1992–1993 apparition.	Early May to late October. Best in early August, rising about 3 hours before the sun. This is a repeat of the 1993 apparition.

misery," for example. To the Maya, predicting the movements of Venus amounted to a form of civil defense.

Elements of the Maya pantheon were adopted by the Aztecs, who rose to prominence in central Mexico after Maya culture had declined. The Aztec god of wind and fertility was *Quetzalcoatl*, who represented the predawn appearance of Venus and whose name also means "feathered serpent." His twin, *Xolotl*, the deity of magicians, shone as the evening star.

Unlike the Babylonians and the Maya, the early Egyptians weren't inclined to systematic observation. Still, they referred to Venus as "the crosser" or "the star which crosses" and symbolized it as *Bennu*, a legendary heron-like bird in which the reincarnated soul of *Osiris* resided. Like another legendary bird, the phoenix, Bennu would die in flames and be reborn from its own ashes — a tantalizing suggestion that the Egyptians knew the true identity of the morning and evening stars. After all, the evening star descends into the flame of the sun only to emerge reborn into the morning sky.

The brilliance of Venus still takes us by surprise. On the astronomical magnitude scale, where smaller numbers indicate greater brightness, Venus hits -4.6 at its best — only the sun and moon are brighter. The planet figures prominently among reports of "unidentified flying objects," and there's even a story about a chagrined air traffic controller who, taking the morning star for an arriving flight expected from the east, radioed Venus its landing instructions. The American physicist J. Robert Oppenheimer related in a 1950 letter to Eleanor Roosevelt a similar case of mistaken identity that occurred in Los Alamos, New Mexico, in the tense days before the test of the first nuclear bomb. He wrote:

I remember one morning when almost the whole project was out of doors staring at a bright object in the sky through glasses, binoculars and whatever else they could find; and nearby Kirtland field reported to us that they had no interceptors which had enabled them to come within range of the object. Our director of personnel was an astronomer and a man of some

Figure 2-2. *Venus conceals its true character beneath a yellow veil of thick, sulfurous clouds. The clouds, featureless in visible light, show detail only in images taken through ultraviolet filters. The U.S. Pioneer Venus spacecraft returned this image in 1980. (Photo by NASA/Ames)*

human wisdom; and he finally came to my office and asked whether we would stop trying to shoot down Venus.

Dominique Arago, the director of the Paris Observatory in the early 1800s, liked to tell of Napoleon Bonaparte's daytime encounter with Venus. On December 10, 1797, while on his way to a victory banquet at Luxembourg Palace in celebration of his successful Italian campaign, Napoleon noticed that the crowds lining the streets were looking at the sky instead of paying attention to the procession. He asked what was going on and was told that a star could be seen shining above the palace. The timely appearance of Napoleon's daytime "star" (Venus, of course) might have given the impression of celestial endorsement.

Our understanding of the physical nature of Venus has come with agonizing slowness. Even though Venus is the brightest and nearest of the planets, observed over four millennia, little was definitively known about the actual conditions on the planet until the middle of the twentieth century. There is a very good reason for the slow progress: Venus is totally enshrouded in a reflective mantle of bright, pale yellow clouds (Fig. 2-2). Those clouds at once reveal the planet to the stargazer and veil it from the inspection of scientists.

Near the part of its orbit that brings it closest to Earth, Venus appears as a brilliant, featureless crescent when viewed through a telescope. The Italian astronomer Galileo

Galilei recognized in 1610 that Venus mimics the moon by going through a cycle of phases. This was hard evidence that the planet revolved around the sun and supported the heliocentric ideas proposed by the Polish monk Nicolaus Copernicus in 1543. The next breakthrough had to wait until the mid-1700s, when astronomers finally found proof of an atmosphere. Yet even by 1960 such basic information as the planet's rotation period remained a lively subject for speculation.

What little scientists did know suggested that Venus was a slightly smaller version of our own planet. While its proximity to the sun gives it twice the sunlight we receive, the planet-wide cloud layer reflects nearly all of it back into space. The clouds give the planet its brilliance, but they also assure that Venus absorbs less solar energy than Earth (and about the same as Mars) despite its nearness to the sun. Some observers believed the thick, featureless cloud deck was formed by water droplets in an atmosphere much like our own. Many thought the length of the Venus day was about the same as Earth's. In short, Venus appeared to be Earth's twin.

A very different picture is evident from Table 2-2, which summarizes the important physical and orbital data now known for Venus. In size and composition Venus does resemble Earth, but there the similarity ends. The temperature on the planet's surface averages 457° C (855° F) day and night — hot enough to melt tin, lead, and zinc. The bulk of the atmosphere consists of carbon dioxide gas, which explains the high surface temperature as "glo-

Table 2-2. Facts About Venus and Mercury

	Venus	Mercury
Diameter:	12,104 kilometers 7,521 miles 95% that of Earth	4,878 kilometers 3,031 miles 38% that of Earth
Surface temperature:	457° C 855° F	167° C 333° F
Surface atmospheric pressure:	90 Earth atmospheres equivalent to the pressure at a depth of 905 meters, or 2,970 feet, under the ocean.	Essentially zero
Atmospheric composition:	96% carbon dioxide 3.5% nitrogen	98% helium 2% hydrogen
Moons:	None	None
Rotation period:	243.01 days The planet spins opposite to the rotation of Earth.	58.65 days
Sidereal orbital period:	224.7 days	87.97 days
Synodic orbital period:	583.92 days	115.88 days
Mean distance from sun:	108.2 million kilometers 67.2 million miles 72.3% that of Earth	57.9 million kilometers 36.0 million miles 38.7% that of Earth
Orbit inclined to Earth's:	3.39 degrees	7.00 degrees

bal warming" gone wild. The bright clouds reflect away all but 25 percent of the sun's energy and only a small fraction of this actually reaches the surface. But the atmosphere makes this small fraction count. The short-wavelength energy (light) reaching the surface is absorbed and then reradiated at longer (infrared) wavelengths — heat. Atmospheric carbon dioxide prevents the heat from easily escaping into space. The gas repeatedly absorbs and reradiates it, delaying its departure. By the time the outgoing heat radiation balances the incoming solar energy, there's already enough heat bouncing between the surface and the atmosphere to keep them both plenty hot. This process, nicknamed the "greenhouse effect," also operates in our atmosphere where water vapor is the most important "greenhouse" gas.

The atmosphere of Venus held still more surprises. In the late 1960s the Soviet Union attempted to land instru-mented capsules on the planet's surface as part of their ongoing Venera series of space missions. Although the first three attempts succeeded in dropping capsules into the atmosphere, each probe ceased operating before reaching the surface. When the next probe survived its descent and returned surface pressure measurements, it became clear why the earlier Venus landers had failed: They were crushed as they descended. The dense atmosphere of Venus presses onto the surface with a force ninety times greater than Earth's, equivalent to the pressure experienced by a submarine at a depth of over half a mile (nearly one kilometer) beneath our oceans.

This catalog of atmospheric horrors would not be complete without mentioning the contents of those perpetual yellow clouds. They aren't made of water droplets because water in any form is scarce on Venus (about 0.01 percent in the atmosphere). Instead the clouds form out of

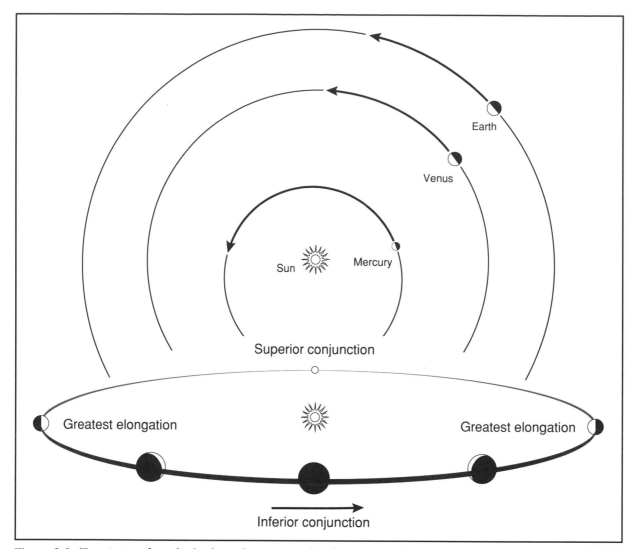

Figure 2-3. *Top: A view from high above the "racetrack" of the inner solar system. Arrows indicate the average distance each planet travels in a month. Both Venus and Mercury move faster than Earth. Bottom: How we see Venus or Mercury depends on their positions relative to Earth. The diagram shows the changes in apparent size and phase both planets undergo during their visibility cycle. They pass between Earth and the sun at inferior conjunction.*

droplets of concentrated sulfuric acid, with traces of hydrochloric and possibly hydrofluoric acids thrown in as well. So, in addition to its stifling heat and crushing pressure, the atmosphere of Venus is also the most corrosive in the solar system.

Venus has seen more terrestrial hardware than any other planet; more than two dozen Soviet and American space missions have visited the planet since 1961. Eleven spacecraft have returned data from the surface, but most survived less than two hours in the hostile environment. Four Soviet landers even transmitted images from their landing sites, revealing a sterile, rocky, arid landscape bathed in a soft orange light. Only about two percent of the incident sunlight reaches the surface, about the same as on a heavily overcast day here on Earth.

The most recent missions to Venus have focused on the detailed mapping of surface features. Scientists have identified impact craters and volcanic landforms, and there is evidence to suggest that Venus may still undergo geologic activity such as earthquakes and volcanoes. The clouds block out visible light, so the mapping of Venus had to await the development of "imaging radar." This instrument pierces the cloudy veil with radio waves to produce photograph-like pictures of the terrain.

It was through Earth-based radar studies in 1961 that scientists were at last able to pin down the rotation rate of Venus. It spins east to west, opposite to the direction of most other planets, and takes 243 Earth days to make one complete rotation. Since Venus completes one orbit around the sun in just 225 Earth days, the "Venus day" is actually longer than the "Venus year". No one knows why Venus spins so slowly.

The View from Earth

Observers had long known that the motions of Venus and Mercury were intimately tied to the sun, but explaining just why this was so proved difficult. To fully appreciate why we see Venus and Mercury more easily at some times than at others, it's necessary to understand just how they move.

If we could view the solar system from a far-flung point high above the sun's north pole, we would see all the planets moving around the sun in their elliptical orbits much like cars on a racetrack (Fig. 2-3). Those in the innermost lanes circle fastest, the outermost the slowest. Turn the solar system on its "side" — that is, view it along the plane of the earth's orbit, the "ground level" of the racetrack — and the planets seem to slide from one side of the sun to the other. From behind the sun each planet moves to the left (east) until it appears farthest from the sun, at which point it doubles back. Proceeding now to the right (westward), the planet's apparent distance from the sun shrinks. It eventually crosses our line of sight to the sun and continues west until it again reaches a maximum angle, reverses course, and heads east to return to the far side of its orbit.

From our perspective on earth we not only view the solar system along the plane of the racetrack, but from one of the innermost lanes. The planets on the outer orbits arc away from the sun and pass behind us, where we can watch them throughout the night. In truth only two planets exhibit the back-and-forth motion described above. That's because only two planets — Mercury and Venus — move on the inside track as seen from Earth's position. For skygazers, both planets put on their best shows near the time of *greatest elongation*, the point where they reach their largest angle east or west of the sun. For Venus this is between forty-five and forty-seven degrees.

With that image in mind, we can track the relative positions of Earth and Venus through one complete cycle of visibility. We'll begin as before, with Venus located on the far side of the sun as seen from Earth at the point termed *superior conjunction*. Venus, now at its greatest distance from us, is lost in the glare of the sun. If the planet could be seen, a telescope would reveal a very small, fully illuminated disk. Venus moves farther east each day, setting progressively later than the sun. After a few weeks we can see it low in the west, glimmering in the twilight of dusk before following the sun below the horizon. It continues sliding eastward and pulling away from the sun, each day appearing higher above the horizon after sunset and becoming more noticeable in the evening sky. When Venus reaches greatest eastern elongation (about seven months after superior conjunction), it dominates the early evening and sets a couple of hours after the sun. To

telescopic observers the planet's disk has grown larger, but now only half of its sunlit side faces Earth.

Venus now runs on the near side of its orbit, approaching us as it reverses course and rushes westward. Its disk continues to grow, but a telescope reveals an ever brighter though slimmer crescent. The planet grows brighter until about five weeks after greatest elongation, when the fading light from its ever-shrinking crescent finally offsets its increasing angular size. Venus falls quickly from its summit in the evening sky, taking just ten weeks to plunge back into the sun's glare. It passes between Earth and the sun (*inferior conjunction*) and gains a lap on us. So ends the evening visibility period, or *apparition*, of Venus.

The planet isn't lost in the sun's radiance for long, though, and it quickly pulls west of the sun and begins its morning apparition. Venus climbs higher into the predawn sky as rapidly as it fell from the evening sky, brightest about five weeks after inferior conjunction and reaching greatest western elongation five weeks after that. The planet then reverses course and, now on the far side of its orbit, begins a lazy, seven-month-long descent toward the horizon. Venus completes the cycle when its slow eastward slide to the sun brings it back to superior conjunction.

Figure 2-4 tracks Venus throughout typical evening (1992-1993) and morning (1996-1997) apparitions and illustrates its changes in position and speed. Table 2-1 lists the dates when Venus can be seen as a conspicuous evening and morning star for observers at mid-northern latitudes. Although Venus is bright enough to be seen even in daylight, we limited the dates in the table to times when Venus appears at least ten degrees above the horizon about forty-five minutes after sunset or before sunrise. Looking for Venus closer to dusk or dawn will extend these dates somewhat. See the appendix for more detailed information.

We said earlier that Venus completes one orbit around the sun in 225 days (its sidereal period), but the planet's visibility cycle takes 584 days, or 1.6 years (its synodic period). The fact that the Earth also moves, that our viewpoint within the solar system's racetrack isn't stationary, explains the difference. (We came across a similar discrepancy in Chapter 1's discussion of the moon.) For the same geometry to recur between Earth, Venus, and the sun, Venus must make up the extra distance traveled by the Earth before reappearing in the same part of our sky.

Although Venus is always rather easy to find, it only commands attention when its greatest elongation occurs in the right season. At its best, Venus rises nearly four hours before dawn or lingers as long after sunset. All of the planets stay close to the ecliptic, which intersects the horizon most steeply near the start of spring and fall. In the northern hemisphere, Venus gives its best evening displays when greatest eastern elongation occurs near the start of spring; the worst when it occurs near the end of summer (Fig. 2-5).

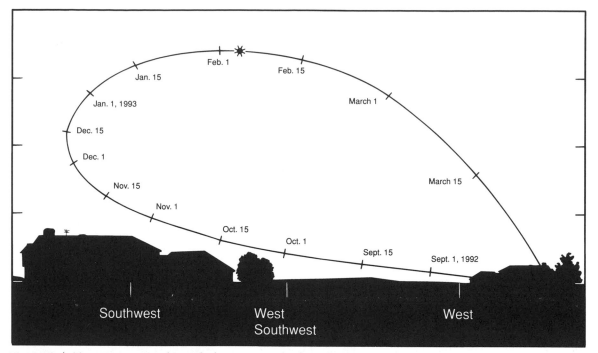

Figure 2-4. *Top: Venus arcs through the evening sky from late August 1992 to late March 1993, as seen forty-five minutes after sunset. The planet reaches its greatest elongation east of the sun on January 19, but keeps climbing above the horizon until February 4. Venus repeats this fine display in 2001. Bottom: Venus tracks through the morning sky from late June 1996 to late January 1997, forty-five minutes before sunrise. The planet attains its greatest angle west of the sun on August 20, but rises to slightly higher altitudes until September 5.*

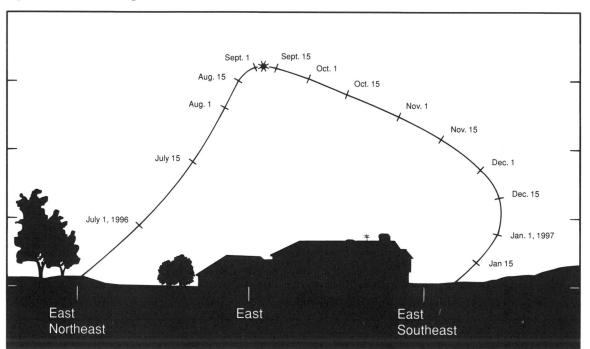

Earth and Venus share a curious relationship, one that contributed to Mayan and Babylonian fascination with Venus. The planet completes five synodic cycles (2,919.6 days) in almost exactly eight years (2,922 days). That is, after every set of five evening and morning apparitions

Venus replays its performance. For example, the planet's fine evening show in the beginning of 1993 repeats almost exactly in 2001. The difference between eight Earth years and five synodic Venus cycles is less than sixty hours. So to update the 1993 diagram in Figure 2-4 for the 2001

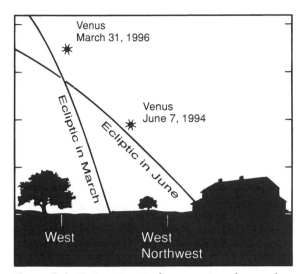

Figure 2-5. *Venus gives its best evening shows when its greatest angle east of the sun occurs near the end of winter, which happens in 1996. The worst case occurs when the planet reaches greatest eastern elongation near the start of summer, as in 1994. The planet is forty-six degrees east of the sun on both occasions; the steepness of the ecliptic accounts for the difference in altitude.*

apparition, just subtract two days from the dates beside the planet's track.

Every century or so, Venus puts on a pair of very special performances when it passes between Earth and the sun. The orbit of Venus does not lie in the same plane as Earth's, so usually Venus passes slightly above or below the sun at inferior conjunction. Rarely, though, it tracks directly across the sun's disk — an event called a *transit*. The planet shows up in silhouette, a small black dot just one-thirtieth the apparent diameter of the sun. For a transit

to occur, Venus must reach inferior conjunction when it lies near one of the points where its orbit cuts the plane of the earth's orbit. So Venus only transits the sun when inferior conjunction takes place within two days of June 7 and December 9. Astronomers observed the first transit of Venus in 1639, and it was during the transit of 1761 that they found evidence of the planet's atmosphere. The next transit occurs on June 8, 2004. The event lasts over six hours and, although favoring the eastern hemisphere, a portion of it is visible from much of the United States. Venus transits the sun again eight years later, on June 5-6, 2012. You'll have to wait till 2117 for the next one.

Mercury: Messenger of the Gods

The Babylonians associated Mercury with their god of wisdom, *Nabu*, and called it by the name *Gud-ud*. The meaning of Mercury's earliest Egyptian name, *Sbg(w)*, is unknown, but a later text refers to the planet as "*Set* in the evening twilight, a god in the morning twilight." Mercury apparently took on the malevolent disposition of Set, brother of Osiris, when it appeared in the evening, but changed identities when it switched to the morning sky. The Greek name *Stilbon* ("twinkling star") seems inspired by the planet's appearance, shimmering in the unsteady air near the horizon. The Greeks associated Mercury with the swift-footed messenger of the gods, *Hermes*, who watched over travelers and brought good fortune in commerce; our name for the planet comes from his Roman counterpart. The Anglo-Saxons named the day ruled by Mercury after their chief deity, *Woden*, from which Wednesday ("Woden's day") derives.

Mercury runs on the solar system's innermost track, completing an orbit around the sun in just eighty-eight days. The telescope reveals little about the planet except

Figure 2-6. *Mariner 10 encountered Mercury three times between 1974 and 1975, photographing about half of the planet's rugged surface. This photomosaic of Mercury combines eighteen separate images and shows about two-thirds of the planet's northern hemisphere. In many respects Mercury resembles an enlarged version of our own moon. (Photo by NASA)*

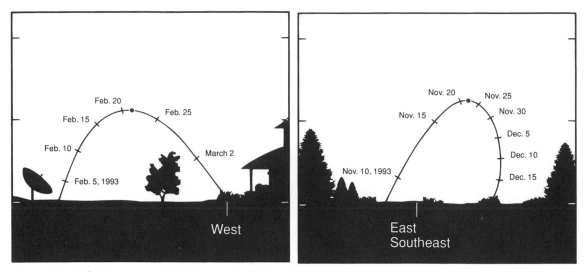

Figure 2-7. *Left: Mercury in the evening sky from February 1 to March 4, 1993, as viewed about thirty minutes after sunset; it reaches greatest elongation on February 21. Right: Mercury's track through the morning sky between November 9 and December 20, 1993, as seen about thirty minutes before sunrise; greatest elongation occurs November 22.*

its cycle of phases. Radar studies in 1965 determined that Mercury spins once on its axis every fifty-nine days, which happens to be exactly two-thirds of its orbital period. This means that Mercury makes three complete rotations for every two orbits around the sun. Table 2-2 lists Mercury's other vital statistics.

A U.S. spacecraft named *Mariner 10* lifted the veil of mystery surrounding Mercury's physical nature. Its cameras photographed about half of the planet's surface during three flybys in 1974 and 1975 (Fig. 2-6). From a distance Mercury could be mistaken for our own moon. One significant difference is the planet-wide system of geologic faults that arose when Mercury's interior cooled and caused its surface to shrink.

Mercury's overall motion through our sky resembles that of Venus, with periods of visibility centered on the dates of its greatest angular distance from the sun. Mercury averages six greatest elongations each year, but they carry the planet no more than twenty-eight degrees from the sun. So while Mercury can rival all but the brightest stars, it never climbs high enough into the sky to emerge from the twilight glow of dusk or dawn (Fig. 2-7).

Table 2-3 summarizes Mercury's best morning and evening apparitions between 1993 and 2001. We've already discussed how the steepness of the ecliptic affects Venus's visibility; this proves even more important for Mercury. A glance at the table shows that Mercury puts on its finest shows as an evening star in the spring and as a morning star in the fall.

Locating Mercury is difficult at best — even Nicolaus Copernicus, the founder of modern astronomy, lamented that it had always given him trouble. But there's no better method of finding Mercury than that of using the moon and planets to point the way. The apparitions illustrated in

Figs. 2-8 through 2-16 show such celestial guideposts, along with each planet's magnitude on the astronomical brightness scale. Two of these apparitions are worthy of particular note. In the evenings of early March 1999, a chain of planets formed by Saturn, Venus, and Jupiter lead on to Mercury. And on the brisk mornings of late October and early November 2001, Mercury approaches to within one degree of dazzling Venus and remains this close for eleven days!

Like Venus, Mercury also transits the sun. By 2001 it will have crossed the sun's disk fourteen times within a hundred years, slightly better than the average of thirteen transits per century. Mercury's transits require the same geometric conditions as those of Venus, but the speedy planet has more chances to meet them. For a transit to occur, Mercury must come to inferior conjunction within three days of May 8 or within five days of November 10. As a result, November transits outnumber those occurring in May.

On the morning of November 7, 1631, the French astronomer Pierre Gassendi became the first to witness one of these events. Since observing transits requires the use of a telescope and a safe technique for viewing the sun, such as projecting its image onto a wall, we won't dwell on them. Figure 2-17 shows the planet's track across the sun during each of the next four transits, two of which occur in the 1990s. On November 6, 1993, Mercury spends about an hour and a half crossing the sun's southwestern edge; observers in eastern and central Asia, Australia, the eastern coast of Africa, and extreme eastern Europe will be able to see at least part of it. (See the Appendix.) The following event, on November 15, 1999, is a bust for most of the world: Mercury transits along the northeastern edge of the sun — and favors Antarctica.

Table 2-3. Mercury's Best, 1993-2001. Each entry gives the range of dates for which an observer at mid-northern latitudes will find Mercury at least ten degrees above the horizon thirty minutes before sunrise or after sunset.

Year	As Evening Star (In the west after sunset)	As Morning Star (In the east before sunrise)
1993	February 17 to 25 May 31 to June 23	November 16 to 30
1994	February 1 to 8 May 15 to June 8	October 31 to November 12
1995	January 18 to 22 April 30 to May 21	October 16 to November 26
1996	April 14 to 30	September 30 to October 7
1997	March 29 to April 12	January 16 to 20 September 14 to 20
1998	March 14 to 24 July 3 to 11	December 28, 1997, to January 9 August 29 to September 4 December 11 to 26
1999	February 26 to March 7 June 11 to July 2	August 12 to 19 November 25 to December 10
2000	February 11 to 18 May 24 to June 16	July 29 to 31 November 9 to 22
2001	January 25 to February 1 May 8 to 31	October 25 to November 4

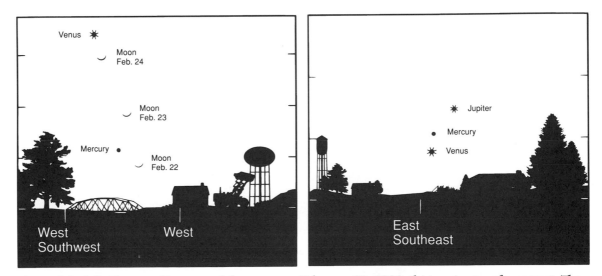

Figure 2-8. *Left: Mercury, Venus, and the moon on February 22, 1993, thirty minutes after sunset. The moon's slender crescent slides past Mercury (-0.3) and then approaches brilliant Venus (-4.6). The moon is here enlarged about four times for clarity. Right: Mercury, Venus, and Jupiter, November 16, 1993, thirty minutes before sunrise. Mercury (+0.3) scoots past Venus (-3.9) as Jupiter (-1.7) climbs out of the morning twilight.*

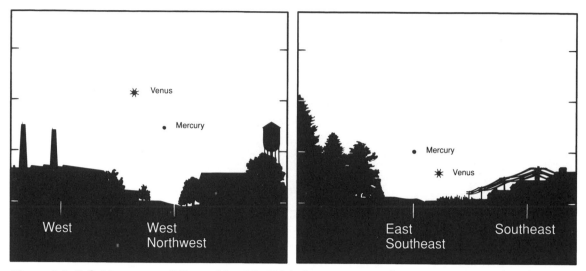

Figure 2-9. *Left: Mercury and Venus, May 27, 1994, thirty minutes after sunset. Mercury (+0.2) chases Venus (-4.0) into the evening sky, making one of its best appearances during the period covered by this book. Right: Mercury and Venus, November 12, 1994, thirty minutes before sunrise. Sinking Mercury (-0.7) passes Venus (-4.3). Binoculars help.*

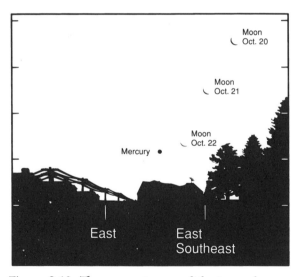

Figure 2-10. *The crescent moon slides toward Mercury (-0.5) between October 20 and 22, 1995, thirty minutes before sunrise. Remember: Tick marks along the sides represent altitude intervals of ten degrees — about the distance covered by your fist at arm's length.*

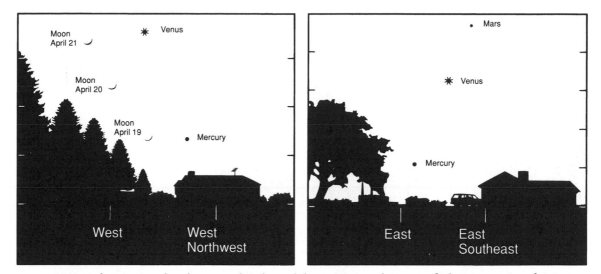

Figure 2-11. *Left: Mercury (-0.1), Venus (-4.5), and the moon, April 21, 1996, thirty minutes after sunset. Right: Mercury (-0.4), Venus (-4.1), and Mars (+1.4), October 3, 1996, thirty minutes before sunrise.*

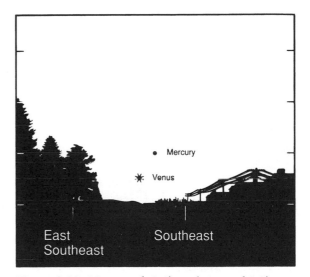

Figure 2-12. *Mercury (+0.0) and Venus (-3.9), very low in the southeast thirty minutes before sunrise on January 18, 1997. Binoculars help.*

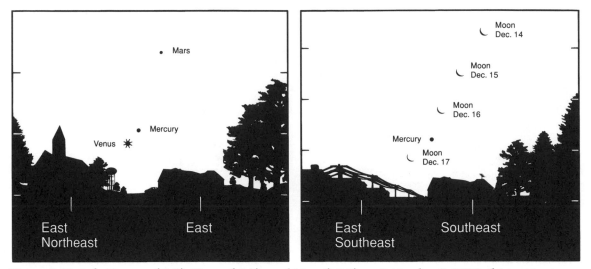

Figure 2-13. *Left: Mercury (-0.3), Venus (-3.9), and Mars (+1.7) on September 1, 1998, thirty minutes before sunrise. Right: Mercury (-0.2) and the waning moon on December 16, 1998, thirty minutes before sunrise.*

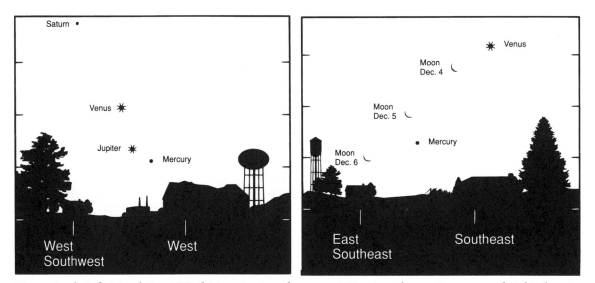

Figure 2-14. *Left: March 3, 1999, thirty minutes after sunset. Here's a chance to see every bright planet but Mars — Saturn (+0.5), Venus (-4.0), Jupiter (-2.1), and Mercury (-0.4). Right: Mercury (-0.5), Venus (-4.2), and the waning moon, thirty minutes before sunrise on December 4, 1999.*

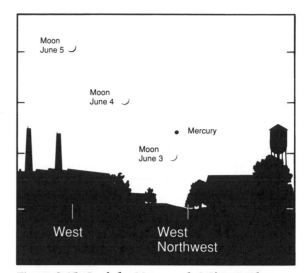

Figure 2-15. *Look for Mercury (+0.2) near the waxing moon on June 4, 2000, about thirty minutes after sunset.*

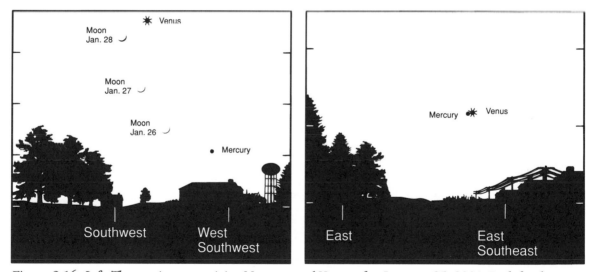

Figure 2-16. *Left: The waxing moon joins Mercury and Venus after January 25, 2001. Look for them about thirty minutes after sunset. The moon is enlarged for clarity. Right: Mercury (-0.5) hovers less than a degree from Venus (-3.9) between October 28 and November 7, 2001. This chart shows the pair on the morning of October 29, when they make their closest approach. Look thirty minutes before sunrise.*

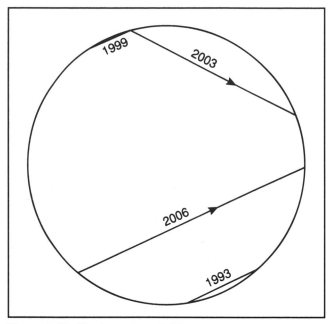

Figure 2-17. *During a transit Mercury moves east to west across the sun's face. This drawing shows Mercury's path across the sun for each of its next four transits. Only the 2006 event favors the western hemisphere.*

3   ECLIPSES OF THE SUN AND MOON

The sun, moon and planets travel through the sky along the same celestial roadway, a track known as the ecliptic. Since these objects move at different speeds, sooner or later a couple of them must arrive at the same place at the same time. The nearer of the two objects will pass in front of the farther one, blocking it from view, or occulting it. The moon frequently occults planets and stars that lie along the ecliptic. Such events are always interesting to observe, but by far the most sensational occultation known to humankind occurs when the moon passes in front of the sun and completely blocks its light, creating a brief, eerie darkness in the middle of the day — a total solar eclipse (Fig. 3-1). The full moon undergoes a similar, though far less spectacular, darkening during a total lunar eclipse. The preeminence of these events explains why the path of the sun, moon, and planets is known as the ecliptic instead of the "occultic."

In folklore, eclipses are most commonly explained as attacks on the sun and moon either by heavenly monsters — dragons in China, snakes in Indonesia, jaguars or wolves in the Americas — or, somewhat more accurately, by one against the other in battle. Some Eskimo groups believed that the eclipsed body had simply left its place to check on earthly affairs. Rio Grande Pueblos feared that when the Sun Father's shield faded he was displeased and had moved away from the earth. In areas of the South Pacific, eclipses were viewed more romantically as the lovemaking of the sun and moon.

Eclipse Basics

For most people, the primary question surrounding any eclipse is simply "Will I be able to see it?" The easiest way to determine whether or not a given eclipse favors your hometown is simply by looking at a map — so this chapter is short on words and long on pictures. The following pages contain thirteen global maps illustrating the best solar and lunar eclipses through 2001. Before discussing these maps, however, it's useful to first understand the hows and whys of eclipses.

A solar eclipse occurs when the new moon passes in front of the sun and covers at least a portion of its disk — we then pass through the moon's shadow. A lunar eclipse occurs when the full moon passes into the shadow cast by the earth, which darkens at least part of the moon. These shadows have two components: a broad, diffuse *penumbra* from which the sun's light is only partially cut off, and a much darker *umbra* within which no part of the sun's disk can be seen.

Since the moon courses through the ecliptic once each month, why don't we enjoy each of these eclipses with the same frequency? Remember that the ecliptic represents the sun's apparent annual path through the sky, or the plane of earth's orbit around the sun. If the moon followed exactly the same path, we'd see a total solar eclipse with every new moon and a total lunar eclipse with every full moon. However, the moon's orbit carries it a bit more than five degrees above and below the ecliptic each month. The moon spends about half the month above the plane and the other half below it. So the new moon usually passes above or below the sun and fails to eclipse it, and the full moon usually passes above or below the earth's shadow and fails to be eclipsed. Eclipses can occur only when a new or full moon lies near the points where its orbit crosses the ecliptic. These two points, called the *nodes* of the moon's orbit, must line up with the sun and the earth in order for an eclipse to take place (Fig. 3-2).

Because the sun, moon, and earth are not themselves points, eclipses can occur even when the nodes aren't exactly aligned. For example, the sun can be eclipsed anytime it lies within 18.5 degrees of one of the moon's

Figure 3-1. *With the solar surface blocked by the moon, the tenuous streamers of the sun's corona became visible during the July 1991 total eclipse. From Mexico the sun remained darkened for nearly seven minutes — the longest span of totality until 2132. (Photo by Bill Sterne)*

nodes. The sun slides along the ecliptic by about one degree each day, so it remains within the "solar eclipse window" for thirty-seven days. Since a new moon occurs every synodic month (29.53 days), at least one solar eclipse is guaranteed every time one of the moon's nodes lines up with the sun.

You might expect that these alignments occur six months apart, once on one side of the sun and once on the other, but this isn't quite true. The nodes of the moon's orbit slide westward along the ecliptic themselves, and this drift causes their alignment with the sun to recur every 173.31 days.

This number plays a role in an important pattern of eclipse occurrence first recognized by the Babylonians sometime before 500 B.C. Suppose we were interested in finding some empirical means to predict eclipses. We know that solar and lunar eclipses depend on a certain phase of the moon, and we know how long it takes before the geometry required for an eclipse recurs. Lunar phases repeat every synodic month, so what we're looking for is a length of time that is divisible both by the synodic month and by the node-alignment interval. As it turns out, 223

synodic months (18 years, 11.3 days) nearly equals the time required for 38 node alignments — the difference is just eleven hours.

This cycle is called the *saros.* It means that any two eclipses, solar or lunar, separated by this length of time share similar characteristics and occur near the same date. For instance, the path of the moon's shadow during the European total solar eclipse of August 11, 1999, makes an almost identical curve across the earth during the eclipse of August 21, 2017. Each succeeding eclipse occurs about eight hours later, so the earth's rotation turns the region that witnessed the preceding eclipse one-third of the way around the globe. As a result, the 1999 eclipse tracks through Europe while the eclipse of 2017 runs through the United States.

The related eclipses separated by one saros cycle make up a saros series, a single strand of similar eclipses running into the past and the future. Because there's an eleven-hour difference between the periods making up the saros, a single series cannot run forever. A typical saros series lasts thirteen centuries and consists of more than seventy eclipses, and when one series ends a new one

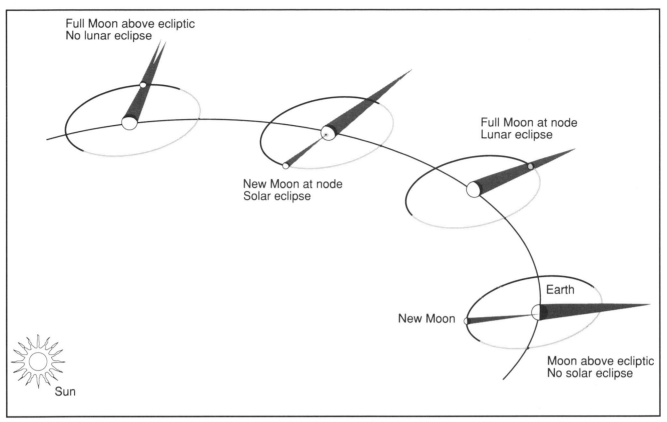

Figure 3-2. *The moon's orbit carries it above (black) and below (light gray) the plane of the earth's orbit around the sun. The moon must lie near that plane, the ecliptic, at new or full moon in order for a solar or lunar eclipse to occur. This happens about once every six months.*

begins. Since there are several eclipses each year, it's obvious that several series run concurrently; about forty are now in progress.

Solar Eclipses

The sun is the nearest star, a self-luminous ball of gas 864,000 miles (1,390,000 kilometers) wide that glows as a result of energy released by nuclear reactions occurring deep within it. The sun's importance to life on earth was recognized everywhere, which made its sudden brief disappearance during a total eclipse all the more frightening. One of the most remarkable coincidences in nature is that the sun and moon both appear the same size as seen from the earth. While they differ in size by 400 times, they also differ in distance by the same factor and both objects appear about half a degree across. If the moon were slightly smaller or a bit farther from us, a total eclipse of the sun — one of the grandest spectacles of nature — would be impossible. Instead, its shadow traces a path across earth's surface within which day briefly turns to night.

There are several types of solar eclipses. If the moon passes over part of the sun's disk but fails to cover it completely, the eclipse is said to be *partial*. Since the sun is never completely covered during a partial eclipse, the moon's dark umbral shadow misses the earth entirely. About 35 percent of all solar eclipses are partial, and up to five partial eclipses may occur in any single year. There are nine partial solar eclipses in the period covered by this book, four of which occur in 2000. Also, a total eclipse may appear partial from your location since totality is visible over such a narrow path.

The moon's umbra may point directly toward the earth but still fail to produce a total eclipse. The orbit of the moon, like those of the planets, is not circular but elliptical. The moon's distance from us varies by more than 14,000 miles (22,000 kilometers) in the course of a month. The moon may appear as much as 10 percent smaller than the sun when it's farthest from earth. So at mid-eclipse, when the new moon lies centered on the sun, observers who would otherwise have seen the sun completely obscured instead see a slender ring, or annulus, of sunlight all around the silhouetted moon. *Annular eclipses*, which make up about 32 percent of all solar eclipses, occur five times in the period this book covers (Fig. 3-3).

A much less common eclipse type represents a transition between the conditions favorable for an annular eclipse and those favoring totality — an *annular total* eclipse. Observers near the middle of the umbra's path see a total eclipse, but those at the beginning and end see an

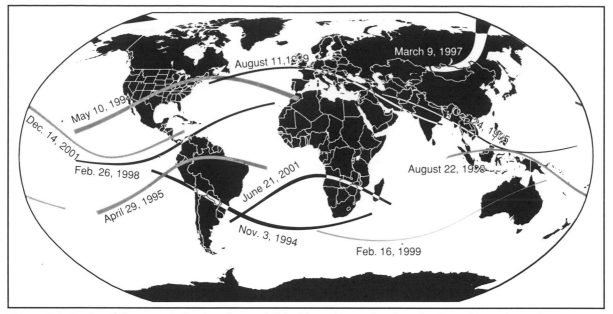

Figure 3-3. *Paths of the moon's shadow for total (black) and annular (gray) solar eclipses through 2001.*

annular eclipse. Only about 5 percent of solar eclipses fall into this category — the next one isn't due until April 8, 2005.

Total eclipses make up the remaining 28 percent. Astronomers estimate that every location on earth experiences totality within 450 years. With total eclipses occurring every eighteen months or so, it's clear that the moon's umbra covers only a very small part of the earth's surface. None of the six total eclipses covered in this book bring totality to North America. So unless you're extremely fortunate and happen to be in the right place at the right time, the only way you're going to see a total eclipse is by planning a trip to follow the moon's shadow. For many people, chasing an eclipse is as good a reason as any to travel — over 100,000 journeyed to Hawaii and Mexico to witness the great total solar eclipse of July 11, 1991 (Figs. 3-4 and 3-5). Hotels were booked years in advance. Why all the fuss? Aside from the unnerving experience of seeing the sun replaced by a black hole in the sky, there are several phenomena that simply cannot be seen during partial or annular eclipses.

The moon's dark umbra is less than 200 miles (320 kilometers) wide and sweeps eastward across the earth's surface at about 1,000 miles (1,609 kilometers) an hour. Those outside this narrow track will see a partial eclipse, but those within it will see the sun's light completely blotted out for up to seven-and-a-half minutes. The moon takes an hour or two to cover the sun. A few minutes before and after totality, when the exposed portion of the sun forms a thin crescent, narrow bands of light and shadow race across the ground. First described in 1706, these *shadow bands* occur because pockets of air refract light from the narrow crescent and winds shift the air pockets.

Then the crescent breaks up into several bright beads of sunlight, portions of the solar disk still shining through deep valleys on the moon's edge. Named for English amateur astronomer Francis Baily, who described their appearance during the annular eclipse of 1836, *Baily's beads* occur for just a few seconds before and after totality. The last bead to fade seems so dazzling that it conjures up the image of a gleaming jewel — it's called the *diamond-ring effect.*

When and only when the diamond ring winks out, it's safe to look at the sun without eye protection. The temperature falls abruptly and the landscape grows about as dark as evening twilight — not exactly "midnight at midday," but dark enough to reveal the brightest stars and planets. A sunset-like glow appears all along the horizon. The moon's shadow covers such a small region that sunlight from the areas outside the umbra tints the sky near the horizon with a reddish glow reminiscent of sunset.

But the big attraction, of course, is the sun. With the brilliant light of its disk completely blocked, faint features normally lost in its glare become visible. Most obvious is the pearly white *corona* that silhouettes the pitch-black moon. This is an extremely hot, tenuous, expanding region of gas about a million times fainter than the sun's surface. Streamers may extend several times the sun's diameter, and observers report seeing details as fine as hair or thread. Pink *prominences* may arch above the moon's rim like tiny tongues of flame. These are huge arcs of hot gas suspended above the sun's surface by intense magnetic fields.

Totality ends with a second diamond ring, quickly followed by another appearance of Baily's beads and then an uninterrupted sliver of sunlight. For a few seconds after

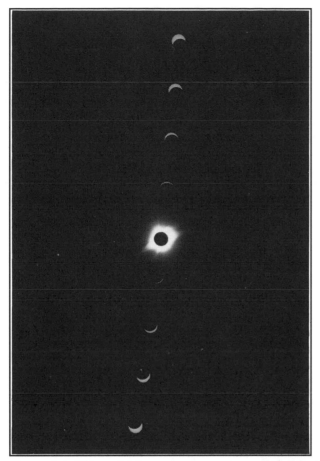

Figure 3-4. *This multiple-exposure sequence of the July 1991 eclipse shows the moon encroaching on the sun, the glow of the corona during totality, and the growing solar crescent as the moon slips off the sun's disk. The partial phases were photographed through a filter that reduced the sun's brightness; the image of totality was a four-second, unfiltered exposure. (Photo by Bill Sterne)*

totality, sites at high altitudes may afford a view of the moon's retreating shadow as it races eastward across the earth's surface.

Humans aren't the only ones who notice the sudden twilight of a total eclipse. Insects seem to react the most dramatically — bees swarm back to their hives, butterflies nestle in the grass, and nocturnal insects emerge as if dusk were really falling. Many bird species roost and try to sleep. Of the mammals studied during various eclipses (including cats, dogs, monkeys, and humans), researchers found no changes in behavior or physiology during totality.

Solar eclipses have played an important role in our current understanding of the sun, but today the unique observing opportunities they provide are available to astronomers by other means. Satellites carrying instruments above the atmosphere can observe the sun in greater detail and over a broader range of its emitted radiation than is possible from the earth's surface. But a satellite-borne instrument costs hundreds of times more than an eclipse trip, and many useful observations simply

don't require that much sensitivity. So while the scientific bonanza from solar eclipses has ended, a few astronomers still chase the moon's shadow.

Viewing the Sun Safely

In addition to visible light, the sun emits infrared and ultraviolet radiation that can seriously damage the light-sensitive cells within the human eye. Even though you may be able to tolerate a prolonged look at the sun during portions of total and annular eclipses, these invisible wavelengths remain dangerous. Materials that decrease the sunlight to tolerable limits but pass the damaging infrared and ultraviolet wavelengths — including sunglasses, smoked glass, and color photographic film — are *extremely dangerous* and should not be used for direct solar viewing. Safe filters include No. 14 welder's glass or Solar-Skreen, an aluminum-coated sheet of plastic. Once the moon covers up the last portion of the sun's disk, you may safely view the eclipse without protection until the sun reappears. An even safer way to view the partial

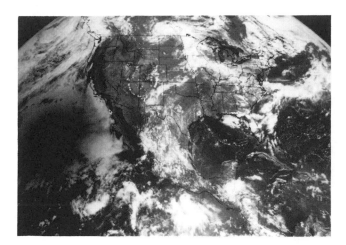

Figure 3-5. *These images, each an hour apart beginning at 18h UT (top), record the July 1991 total solar eclipse from a very different perspective — that of the GOES weather satellite. The moon's shadow sweeps northwest to southeast across the earth's surface at over 1,000 miles (1,609 kilometers) per hour.*
(Photos by Arshag Odabachian, Hewlett-Packard Co.)

phases of an eclipse is to make a pinhole camera from a large box and project the sun's image onto a sheet of paper.

The Solar Eclipse Maps

Figures 3-6 through 3-10 illustrate five of the twenty solar eclipses that occur in the period this book covers. The saddle-shaped grid on each map encloses the portion of the earth's surface touched by the moon's penumbra. All locations within this region experience at least a partial eclipse.

The easiest way to discuss the appearance of a solar eclipse is to think in terms of the percentage of the sun's diameter covered by the moon — so a 50 percent eclipse would mean that at maximum the moon slides halfway onto the sun's disk. Shaded curves on the grid represent the eclipse percentage at 20 percent increments, and cartoons on those curves show roughly how the sun looks at maximum eclipse. For example, take a look at the chart for the May 10, 1994, annular eclipse. The southwestern and central United States and eastern Canada lie within the first curve of the grid — they see the sun between 0 and 20 percent eclipsed at the height of the event. Areas under the next band up see the sun between 20 and 40 percent eclipsed, and so on. For a total or annular eclipse, the path of totality or annularity is represented as a black curve.

The grid also contains two great loops, one on the left and one on the right. The outermost line of the left (western) loop passes through all the locations where the eclipse begins at sunrise. Next comes the line of "maximum eclipse at sunrise," which is where the shaded curves begin, and then the inside line of the loop, which means "eclipse ends at sunrise." Similarly, the lines of the eastern loop represent, from left to right, "eclipse ends at sunset," "maximum eclipse at sunset," and "eclipse begins at sunset."

These maps don't show the time when maximum eclipse takes place at each location. Since no two places on earth see a solar eclipse in exactly the same way at exactly the same time, a table of local circumstances for selected cities accompanies each map. The eclipse predictions in these tables were supplied by astronomer Fred Espenak of the Goddard Space Flight Center in Greenbelt, Maryland. Each table gives the time of maximum eclipse (in Universal Time; see the Appendix for an explanation) and the eclipse percentage for each city; when applicable, the duration of totality or annularity is also given.

Lunar Eclipses

Just as the moon casts a shadow through which the earth occasionally passes, the earth casts a shadow the moon occasionally enters. Lunar eclipses are not nearly as im-

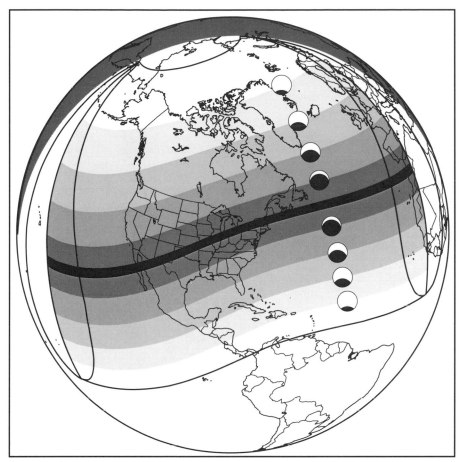

Figure 3-6. *Annular solar eclipse, May 10, 1994. This map is centered near Toledo, Ohio, where the moon covers 94.3 percent of the sun's disk, annularity lasts about 6.2 minutes, and the eclipse track is 143 miles (230 kilometers) wide.*

pressive as their solar counterparts, but they have a special eerie beauty that more than rewards the small effort required to see them. At any one instant a lunar eclipse can be seen by the entire night side of the earth.

At the moon's average distance, the earth's umbral shadow is over two-and-a-half times the moon's diameter. The moon crosses the shadow moving from west to east, the exact path varying greatly from one eclipse to the other. A total lunar eclipse occurs when the moon's entire disk enters into the umbra, otherwise it's a partial eclipse. The earth's shadow also has a faint penumbral component, and sometimes the moon passes only through this region and misses the umbra. These penumbral eclipses often go unnoticed — as do the penumbral phases of partial and total eclipses — and for this reason we won't dwell on them. During the period covered by this book there are eight total, five partial, and seven penumbral lunar eclipses.

A total lunar eclipse may last nearly six hours from start to finish. For the first hour the moon traverses the penumbra, but no darkening is noticeable until nearly the end of this time. The moon then contacts the umbra, and for the next hour a circular shadow creeps across the disk

— it looks almost black by contrast with the part of the moon still in sunlight. At the end of the second hour, the moon's disk lies completely in shadow. While a total eclipse of the sun lasts no longer than seven-and-a-half minutes, the moon can remain completely within the earth's umbra for up to 105 minutes. The eclipse officially ends about two hours after the moon begins its reemergence — one hour to completely leave the umbra, another to exit the penumbra (Fig. 3-11).

If the earth had no atmosphere, the moon would simply disappear from view while in the umbra. The earth's atmosphere acts like a kind of lens, bending sunlight around the edge of our planet and into the umbra. Longer wavelengths of light (orange, red) penetrate the atmosphere better than short wavelengths (blue, indigo), so during totality the moon is faintly illuminated by reddish light. A lunar observer would see the earth surrounded by a thin ring of sunset colors. Just how red the moon looks depends mainly on the amount of volcanic dust and aerosols in the atmosphere. Significant eruptions, such as that of El Chichon in 1982 and Mt. Pinatubo in 1991, can so darken the umbra that a totally eclipsed moon may be almost invisible to the unaided eye.

Local Circumstances for May 10, 1994, Annular Solar Eclipse			
Geographic location	Maximum eclipse (UT)	Percentage of sun eclipsed by moon	Central duration
Los Angeles, CA	16:01	79	
Phoenix, AZ	16:06	87	
El Paso, TX	16:10	94	5:40
Santa Fe, MN	16:18	90	
Denver, CO	16:28	82	
Oklahoma City, OK	16:32	94	3:22
Wichita, KS	16:37	94	2:51
New Orleans, LA	16:38	73	
Little Rock, AR	16:42	87	
Havana, Cuba	16:48	46	
Memphis, TN	16:48	86	
Jefferson City, MO	16:49	94	6:02
Birmingham, AL	16:53	79	
St. Louis, MO	16:53	94	3:37
Springfield, IL	16:57	94	6:11
Winnipeg, MA, Canada	16:57	68	
Atlanta, GA	17:00	77	
Chicago, IL	17:04	94	
Detroit, MI	17:15	94	5:28
Cleveland, OH	17:17	94	4:55
Pittsburgh, PA	17:20	92	
Toronto, ON, Canada	17:24	94	2:43
Buffalo, NY	17:25	94	6:12
Washington, DC	17:26	86	
Syracuse, NY	17:31	94	6:02
New York, NY	17:35	89	
Montpelier, VT	17:40	94	5:36
Concord, NH	17:42	94	4:24
Boston, MA	17:43	93	
Augusta, ME	17:46	94	6:01
Halifax, NS, Canada	17:59	94	5:53
London, England	18:36	54	
Madrid, Spain	18:51	78	
Lisbon, Portugal	18:54	86	

The Lunar Eclipse Maps

Figures 3-12 through 3-19 illustrate the best total and partial lunar eclipses visible from the U.S. through 2001. There are two diagrams for each eclipse. The top drawing simply shows the path of the moon through the earth's penumbral and umbral shadows and the Universal Time for its contacts. The bottom drawing is a world map that shows the regions on the earth where the umbral stages of an eclipse will be visible. For example, consider the total eclipse of June 4, 1993. Observers throughout the Pacific Ocean enjoy a full view of the eclipse's total phase (dark shading). Regions such as California and India see the moon set or rise during totality (medium shading). For the least favored areas, including the central United States, Mexico, and Madagascar, no portion of the total eclipse is seen — the moon sets or rises during the partial phases that precede and follow totality.

Local Circumstances for February 26, 1998, Total Solar Eclipse

Geographic location	Maximum eclipse (UT)	Percentage of sun eclipsed by moon	Central duration
Mexico City, Mexico	17:05	42	
Guatemala City, Guatemala	17:22	64	
New Orleans, LA	17:43	27	
Panama City, Panama	17:43	95	
Bogota, Colombia	17:51	88	
Havana, Cuba	17:54	53	
Atlanta, GA	17:59	25	
Miami, FL	18:02	50	
Kingston, Jamaica	18:02	75	
Nassau, Bahamas	18:08	55	
Maracaibo, Venezuela	18:05	104	2:52
Oranjestad, Aruba	18:11	104	2:50
Willemstad, Netherlands Ant.	18:13	104	1:60
Caracas, Venezuela	18:16	93	
Basse-Terre, Guadeloupe	18:32	104	1:20
St. Johns, Antigua	18:32	104	2:09
Bridgetown, Barbados	18:33	89	

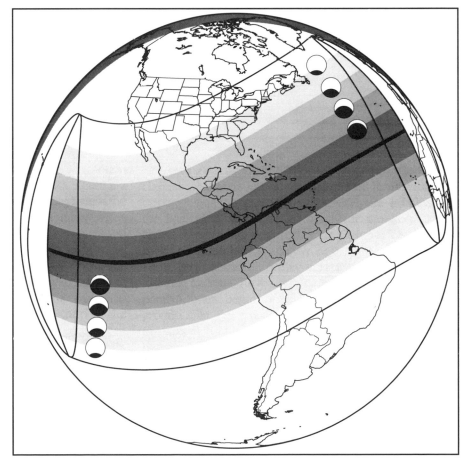

Figure 3-7. *Total solar eclipse, February 26, 1998. Beginning in the equatorial Pacific at sunrise, the moon's shadow sweeps over southeastern Panama, and northeastern Colombia and Venezuela around midday. The eclipse path is no more than 94 miles (151 kilometers) wide, with totality lasting about 4.1 minutes at best.*

Local Circumstances for August 11, 1999, Total Solar Eclipse

Geographic location	Maximum eclipse (UT)	Percentage of sun eclipsed by moon	Central duration
Dublin, Ireland	10:12	93	
London, England	10:20	97	
Paris, France	10:23	99	
Algiers, Algeria	10:24	63	
Antwerp, Belgium	10:26	96	
Amsterdam, Netherlands	10:27	93	
Luxembourg	10:29	102	1:18
Bern, Switzerland	10:31	96	
Milan, Italy	10:35	92	
Copenhagen, Denmark	10:38	82	
Berlin, Germany	10:40	89	
Prague, Czechoslovakia	10:42	95	
Rome, Italy	10:43	84	
Stockholm, Sweden	10:44	70	
Vienna, Austria	10:46	99	
Budapest, Hungary	10:52	99	
Belgrade, Yugoslavia	10:56	98	
Sofia, Bulgaria	11:04	94	
Bucharest, Romania	11:07	103	2:23
Moscow, Russia	11:14	63	
Jerusalem, Israel	11:43	79	
Riyadh, Saudi Arabia	12:12	77	
New Delhi, India	12:26	87	
Karachi, Pakistan	12:27	102	1:13

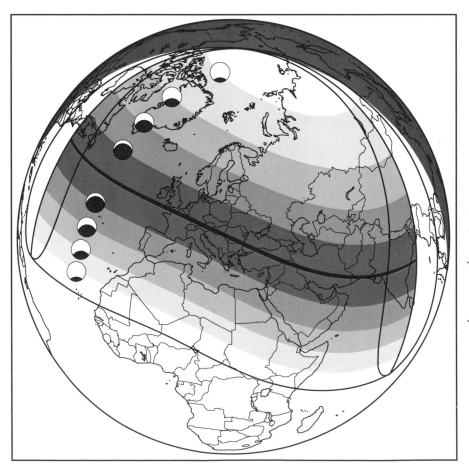

Figure 3-8. *Total solar eclipse, August 11, 1999. The rising sun over New Brunswick and New-foundland will be deeply eclipsed, with the path of totality striking earth just south of land. Plymouth, England, is the largest city near the first landfall of the moon's umbra; it then passes over northern France and through Germany, Austria and Hungary. The map's center is near Bucharest, Romania, where totality lasts 2.4 minutes. The moon's shadow, 70 miles (112 kilometers) at its widest, sweeps over Turkey, Iran, southern Pakistan, and across India before leaving earth.*

Local Circumstances for December 25, 2000, Partial Solar Eclipse

Geographic location	Maximum eclipse (UT)	Percentage of sun eclipsed by moon
San Francisco, CA	16:21	18
Seattle, WA	16:29	37
Phoenix, AZ	16:30	21
Calgary, AB, Canada	16:41	49
Denver, CO	16:44	40
Edmonton, AB, Canada	16:44	53
Houston, TX	17:00	31
Winnipeg, MA, Canada	17:02	59
New Orleans, LA	17:12	35
Milwaukee, WI	17:16	56
Nashville, TN	17:19	47
Cleveland, OH	17:30	56
Havana, Cuba	17:33	26
Pittsburgh, PA	17:34	55
Toronto, ON, Canada	17:34	59
Miami, FL	17:38	32
Washington, DC	17:41	54
Ottawa, ON, Canada	17:41	61
Montreal, QC, Canada	17:45	61
New York, NY	17:47	56
Boston, MA	17:52	58
Halifax, NS, Canada	18:05	58

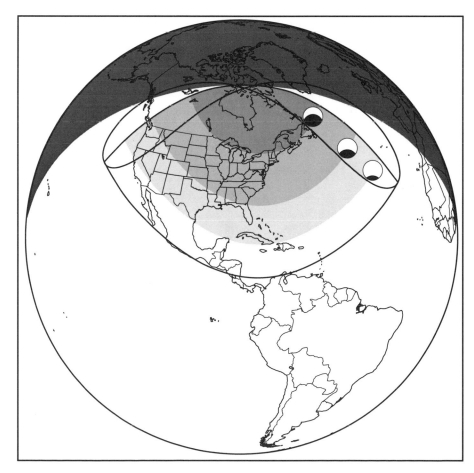

Figure 3-9. *Partial solar eclipse, December 25, 2000. A partially eclipsed sun rises over Vancouver, and sets over Newfoundland, with most of Canada, the United States, and Mexico seeing a significant partial eclipse. The event is best north of Hudson Bay.*

Local Circumstances for December 14, 2001, Annular Solar Eclipse

Geographic location	Maximum eclipse (UT)	Percentage of sun eclipsed by moon	Central duration
Honolulu, HI	19:27	84	
St. Louis, MO	22:02	26	
Indianapolis, IN	22:04	28	
Monterrey, Mexico	22:05	44	
Little Rock, AR	22:05	32	
Frankfort, KY	22:07	32	
Houston, TX	22:08	40	
Nashville, TN	22:09	34	
Jackson, MS	22:11	39	
Chattanooga, TN	22:11	37	
Mexico City, Mexico	22:13	59	
Greenville, SC	22:14	41	
New Orleans, LA	22:14	44	
Atlanta, GA	22:14	41	
Orlando, FL	22:22	56	
Miami, FL	22:25	63	
Havana, Cuba	22:26	67	
Guatemala City, Guatemala	22:26	80	
Belize City, Belize	22:27	75	
San Salvador, El Salvador	22:28	84	
Tegucigalpa, Honduras	22:29	85	
Managua, Nicaragua	22:31	91	
San Jose, Costa Rica	22:33	96	1:06
Colon, Panama	22:35	89	
Barranquilla, Colombia	22:35	87	

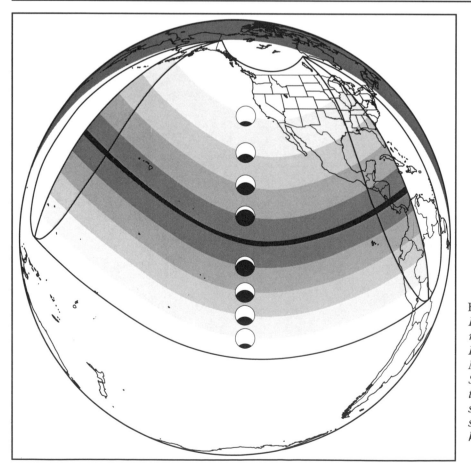

Figure 3-10. *Annular solar eclipse, December 14, 2001. Annularity makes a landfall only over Costa Rica and extreme southeast Nicaragua. Most of the United States sees only a slight partial in the afternoon. Annularity is just short of four minutes long for seagoing observers in the 125-kilometer-wide path at the equator.*

Figure 3-11. *These photographs, taken at intervals of ten minutes, follow the moon as it leaves earth's umbra (beginning at right) during the August 1989 total lunar eclipse. (Photo by Bill Sterne)*

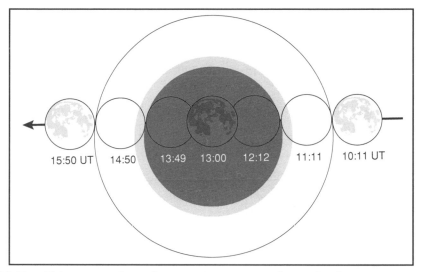

Figure 3-12. *Total lunar eclipse, June 4, 1993. Top: This cartoon shows the moon's position in the earth's faint penumbra (outer circle) and dark umbra (inner circle) at different phases of the eclipse, with times in Universal Time. Below: The Pacific, including Hawaii and Australia, sees the moon's entire passage through the earth's umbra. Along the western coast of North America the moon sets after totality begins, and those a bit farther east can see the beginning partial eclipse before moonset.*

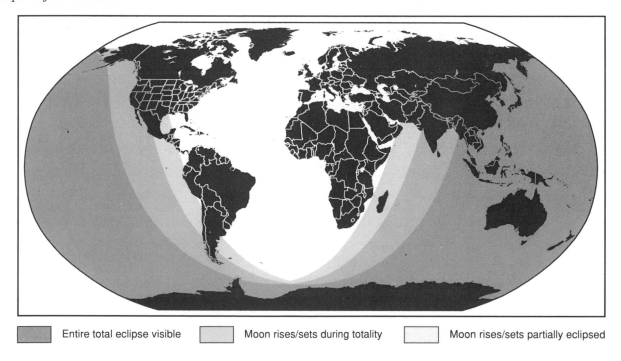

Entire total eclipse visible   Moon rises/sets during totality   Moon rises/sets partially eclipsed

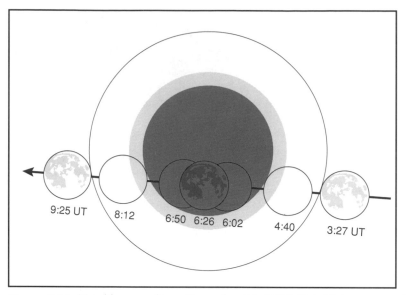

Figure 3-13. *Total lunar eclipse, November 29, 1993. Top: This cartoon shows the moon's position in the earth's faint penumbra (outer circle) and dark umbra (inner circle) at different phases of the eclipse, with times in Universal Time. Below: All of the Americas witness the entire total portion of this eclipse. Will you be able to see the penumbral shading on the moon before it contacts the umbra? How dark will the eclipsed moon appear? Get out and take a look at this one!*

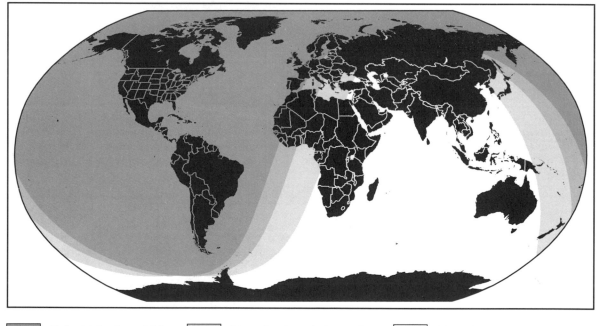

Entire total eclipse visible   Moon rises/sets during totality   Moon rises/sets partially eclipsed

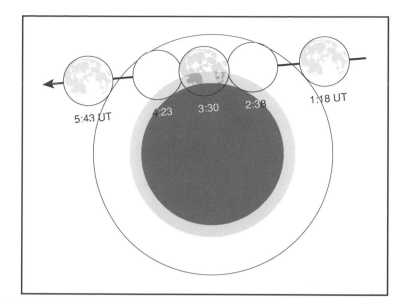

Figure 3-14. *Partial lunar eclipse, May 25, 1994. Top: This cartoon shows the moon's position in the earth's faint penumbra (outer circle) and dark umbra (inner circle) at different phases of the eclipse, with times in Universal Time. Below: The eastern two-thirds of the United States, along with all of Central and South America, will be able to watch the umbral "bite" wax and wane on the moon's southern edge; for the northwestern United States moonrise occurs after the moon enters the umbra.*

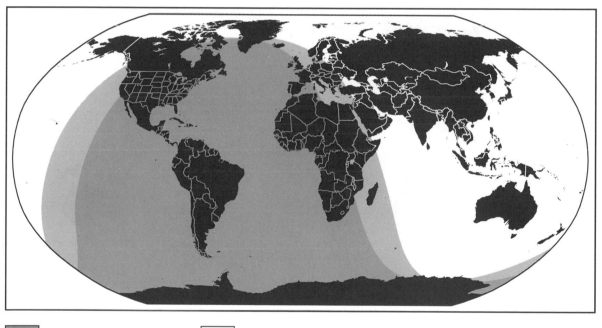

■ Entire umbral eclipse visible   □ Moon rises/sets during umbral eclipse

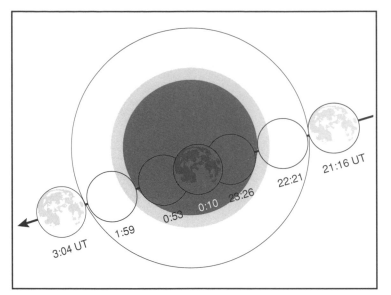

Figure 3-15. *Total lunar eclipse, April 4, 1996. Top: This cartoon shows the moon's position in the earth's faint penumbra (outer circle) and dark umbra (inner circle) at different phases of the eclipse, with times in Universal Time. Below: For the eastern third of the United States, the moon rises during total eclipse; observers in the central United States will see the moon rise partially eclipsed. The entire total portion of the eclipse can be seen from Africa, South America, and Europe.*

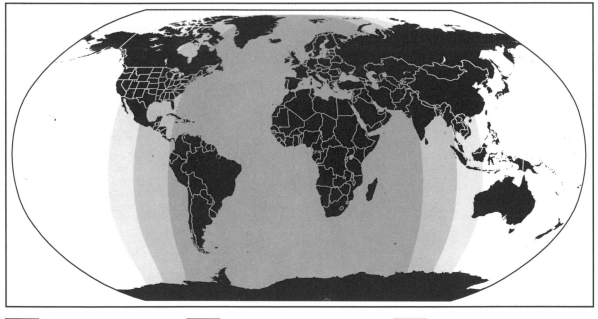

■ Entire total eclipse visible   ■ Moon rises/sets during totality   □ Moon rises/sets partially eclipsed

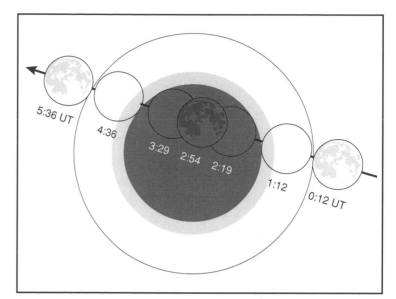

Figure 3-16. *Total lunar eclipse, September 27, 1996. Top: This cartoon shows the moon's position in the earth's faint penumbra (outer circle) and dark umbra (inner circle) at different phases of the eclipse, with times in Universal Time. Below: The moon can be seen throughout the total phase of the eclipse from most of North America, all of Central and South America, Europe, and much of Africa.*

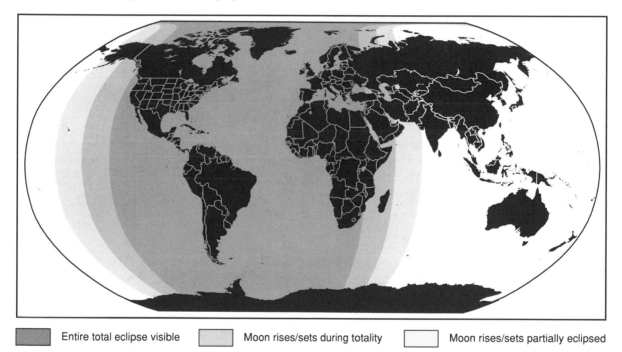

| Entire total eclipse visible | Moon rises/sets during totality | Moon rises/sets partially eclipsed |

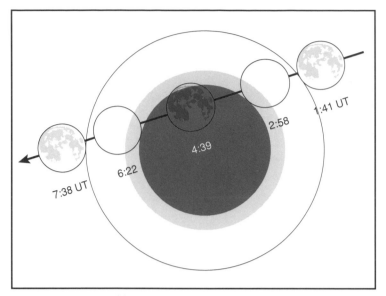

Figure 3-17. *Partial lunar eclipse, March 24, 1997. Top: This cartoon shows the moon's position in the earth's faint penumbra (outer circle) and dark umbra (inner circle) at different phases of the eclipse, with times in Universal Time. Below: Most of North America enjoys a good view of this all-but-total eclipse; Alaska and Hawaii see moonrise after the eclipse is under way.*

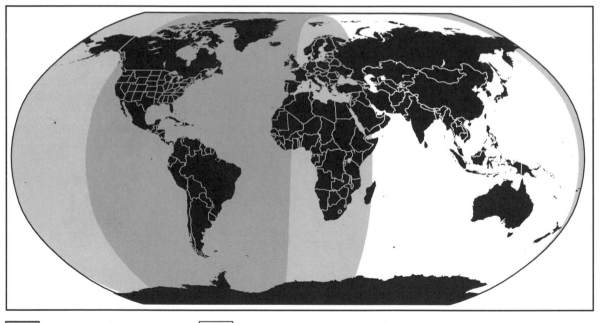

Entire umbral eclipse visible   Moon rises/sets during umbral eclipse

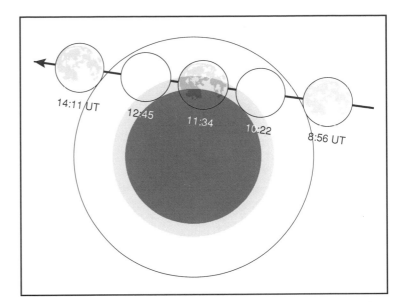

Figure 3-18. Partial lunar eclipse, July 28, 1999. Top: This cartoon shows the moon's position in the earth's faint penumbra (outer circle) and dark umbra (inner circle) at different phases of the eclipse, with times in Universal Time. Below: Much of North America will see the beginnings of the umbral eclipse shortly before moonset; the west coast of the United States will see the moon's entire umbral passage.

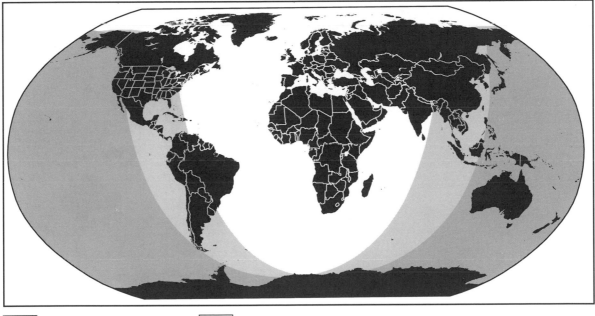

Entire umbral eclipse visible   Moon rises/sets during umbral eclipse

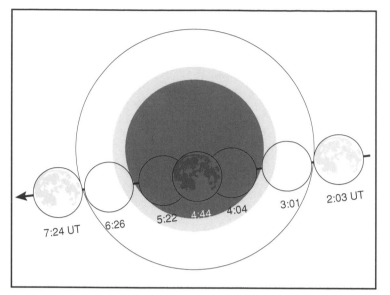

Figure 3-19. *Total lunar eclipse, January 21, 2000. Top: This cartoon shows the moon's position in the earth's faint penumbra (outer circle) and dark umbra (inner circle) at different phases of the eclipse, with times in Universal Time. Below: All of the Americas see the entire total phase of the eclipse. Get out and observe this one!*

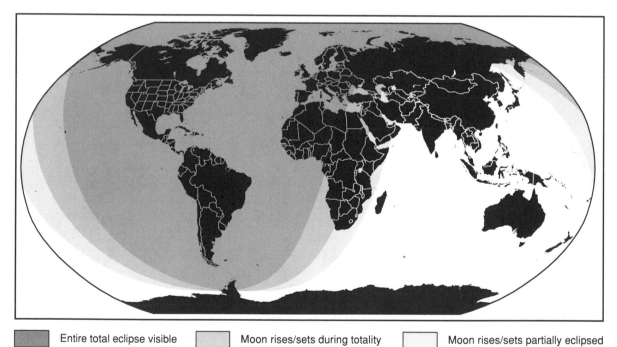

Entire total eclipse visible ▢ Moon rises/sets during totality ▢ Moon rises/sets partially eclipsed

4 MARS: THE RED WANDERER

Mention the word planet to most people, and chances are Mars is the one that pops into their minds. Nowadays, the reason is probably the huge role Mars has played in popular science-fiction views of the planets. (The second word that pops into most peoples' minds is "Martian.") But long before there were "B" movies, science-fiction, and even telescopes, Mars had established itself in the human imagination.

Mars is the archetypical planet. For those who are familiar with the starry sky, this planet's distinctive color and dramatic motions make it the most outstanding "wanderer." While not a true crimson, the hue of the Red Planet is far redder than any other celestial object visible to the naked eye. Near the time when Mars shines brightest, it takes a seemingly chaotic whirl through the heavens. In fact, every planet makes this "retrograde loop," but the one made by Mars is by far the most obvious and dramatic.

Mars soon rewards even its most casual observers. Dawn or dusk are especially good times to appreciate the planet's color, which is set off beautifully against the blue of the twilight sky. When you first notice this faint orange "star" in the morning sky, it seems anything but a wanderer, lingering for weeks just above the eastern horizon. But if you watch the background stars over a week or more, you'll soon see that they are progressing noticeably westward each week, and Mars is indeed traveling eastward through the constellations. If you spot it in the east around 10 P.M., it will be moving very slowly, or not at all, through the background stars. Should you spot it shining brightly to the south at midnight, a week's observations will reveal that it is moving *backward*— westward through the stars. And when it appears in the west after sunset, Mars again lingers in the sky as the background stars slip toward the sun. If you were to observe the planet from its first visibility in the east at sunrise to the time it fades into the western twilight, you'd see it course through two-thirds of the constellations of the zodiac. This is first-class wandering! No wonder Mars has captivated the human mind over the millennia.

Since Mars runs on an outer lane of the solar system racetrack, Earth approaches and overtakes it. Every two years or so we pass between Mars and the sun, an event called *opposition*. This is the middle of the time when Mars is at its best, often brighter than any star in the sky. Take a few moments to get acquainted with this fascinating planet. Table 4-1 lists the dates of upcoming oppositions, when Mars-viewing is best. The illustrations detail the planet's retrograde travels and its most interesting gatherings with the moon and planets. Use the Appendix to determine when Mars next appears. Be sure to compare the planet to the neighboring stars.

The Ancient View

Certainly a star with the idiosyncratic wanderings of Mars fascinated ancient peoples, who set great store in the aspect of the heavens. Many peoples of the Fertile Crescent and the classical world associated Mars with their god of war: *Nergal* of the Babylonians, *Ares* of the Greeks, and of course, the Roman Mars. It's fitting that Mars was considered the father of Romulus and Remus, the mythical founders of the city that went on to conquer nearly all of the western civilization of its day. The first month of the Roman calendar, March, was of course named for the war god. While we're on the subject of calendars, the day of Mars — Tuesday — gets its name from the Anglo-Saxon war god, *Tiw*.

It's often said that this association with war stems from the planet's "bloody" appearance. Indeed, such references are common in ancient astrological texts like this Chinese omen dating from the T'ang dynasty: "Sparkling

Table 4-1. Oppositions of Mars, 1993-2001			
Opposition date	Mag.	Nearest to Earth	Distance from Earth when closest
Jan. 7, 1993	-1.5	Jan. 3	58.2 million miles 93.6 million kilometers
Feb. 12, 1995	-1.2	Feb. 11	62.8 million miles 101.1 million kilometers
March 17, 1997	-1.3	March 20	61.3 million miles 98.6 million kilometers
April 24, 1999	-1.7	May 1	53.8 million miles 86.5 million kilometers
June 13, 2001	-2.4	June 21	41.8 million miles 67.3 million kilometers

Deluder entered Southern Dipper [eastern Sagittarius]. Its color was like blood." While the planet can occasionally appear redder than usual — for example, when it appears near the horizon — we've noted that the Red Planet is rather more orange than crimson. If color is an imperfect explanation, perhaps the planet's departure from the order of the heavens explains its identification with war, which is itself a radical departure from the ordinary rhythm of life.

Not all cultures had such a sanguine view of Mars. The Egyptians called it Horus the Red, after a heroic god akin to the Greek Hercules, and referred to it, aptly

enough, as "he who travels backwards." In China it was *Ying hu*, the "Dazzling Deluder," or "Fire Star," seen as a portent of a variety of troubles, including war. But the Chinese, who associated Venus with war, saw Mars as a force of nature with judicial functions: "Sparkling Deluder is the Master of the Proprieties; when the proprieties are misdone, then the punishment issues from it." In the T'ang dynasty, Mars had names like "the Star of Punishment" and "Holder to the Law," suggesting just the opposite of the chaos of war.

There is some evidence that the Maya of Mesoamerica maintained an interest in the Red Planet. The Dresden

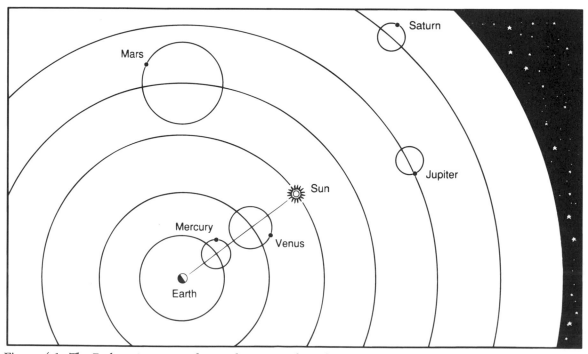

Figure 4-1. *The Ptolemaic system of epicycles can explain the motions of Mars to a high degree of accuracy. Carried along on its orbit, the planet also revolves on a separate circle called an epicycle. It appears to move backward when it passes through the half of the epicycle nearest Earth.*

Codex, which contains the Venus table, also records an enigmatic table that some scholars suspect has its roots in the synodic period of Mars.

Among the Skidi Pawnee Indians of southeastern Nebraska Mars was associated with Morning Star, who created the sun to provide heat and light. Morning Star courted Evening Star — probably associated with Venus — and overcame every hurdle she placed in his path. Through their union, Morning Star and Evening Star brought about the peopling of Earth. Every few years, when the need to appease Morning Star was great, the Skidi Pawnee captured a maiden from a neighboring tribe. Her captor dedicated her to Morning Star by speaking his name — *Opirikuts* — and, after months of ceremonies, she was sacrificed to him to ensure the fertility of the land. Although earthly needs were probably most important in the timing of the sacrifice, the celestial wanderings of Mars appear to have played a role as well.

Mars and the Modern Mind

While the Red Planet has lost its function as a portent of disaster, Mars continues to occupy a special niche in the popular concept of the cosmos. That's appropriate, since besides giving us Martians to fret about at Saturday matinees, the planet has played a pivotal role in the history of astronomy and astronomical speculation. When stargazers began to attempt descriptions of heavenly motion, Mars turned out to be a dazzling deluder indeed. The planet was troublesome for early cosmologists who tried to perceive the order of the planets. Mars's motions ultimately provided the key to our understanding of the mechanics of the solar system.

While the wanderings of the planets were widely noted and tabulated for centuries, it was the Greeks who first offered theoretical explanations for their movements. The earliest such system, devised by Eudoxus in the fifth century B.C., placed the planets on concentric spheres surrounding the earth. The sun and moon each required three moving spheres; the planets required four apiece. Although this system, later refined by Aristotle, gives good approximations for the motions of the sun, moon, Mercury, and Venus, it is less successful with the outer planets. It fails completely for Mars (Fig. 4-1).

The problem has three parts. First, the planets do not move at constant rates; second, they vary in brightness; finally, they actually change direction occasionally, temporarily plying a westward, or retrograde, path before resuming their eastward motion. In the case of Mars, these variations are quite dramatic. Moreover, each of the planet's retrograde loops is quite distinct from the ones that precede and follow it — as you can see for yourself with a glance at Figures 4-5 through 4-7, which chart its five upcoming zigzags.

A few astronomers of the era, notably Aristarchos of Samos who lived around 280 B.C. and the Babylonian Seleucus, held the radical view that the sun lay at the center of the universe. Such a system was, of course, ultimately required to explain the motions of the planets. Two thousand years ago, however, it was intellectually repugnant to most people. And it wasn't sufficiently refined to deal with increasingly accurate observations of the heavenly motions.

A system that did enjoy success was the purely practical system of epicycles. Its three main proponents, Apollonius (230 B.C.), Hipparchus (130 B.C.) and Ptolemy (A.D. 140), were trying less to explain the physical nature of the cosmos than to merely compute the positions of the planets. On purely philosophical (or theological) considerations, these astronomers could allow only circular motions of a constant velocity. This required a large number of modifications, until the system became nearly incomprehensible. But the success of its predictions, along with the demise of the Roman empire and the ensuing "dark ages," combined to keep Ptolemy's system the only acceptable one for over a millennium.

The death knell for the Ptolemaic system rang in 1543 with the publication of *De revolutionibus orbium coelestrium* (*On the Revolution of the Heavenly Spheres*), a small book by a Polish cleric named Mikolaj Kopernigk. Copernicus, as we know him today, began his investigations when he saw that the standard tables of planetary positions, produced by the Ptolemaic method, contained significant errors. He set out to produce better tables. He first attempted to further refine Ptolemy's work, but eventually concluded it was better to scrap the old system. He proposed to replace it with a system similar to the one suggested by Aristarchos nearly two thousand years before. Copernicus, however, also attempted to maintain uniform circular motion. This led to complications and inaccuracies; these, along with theological objections, postponed the universal acceptance of Copernicus's new heliocentric order.

One adherent of the new system was Johannes Kepler, an eccentric (not to say geocentric) assistant to the most famous astronomer of the age, Tycho Brahe. Tycho rejected Copernicus's ideas, but made observations which freed him from the slavish acceptance of classical astronomy that bedeviled many scientists. Among Tycho's best traits was his passion for accurate observation, and under the patronage of Denmark's King Frederick II he built a sophisticated observatory on the island of Ven. He tabulated thousands of planetary positions, using excellent instruments of his own design. Even before their meeting in 1600, Kepler realized the importance of Tycho's observations to any theory of planetary motion. It was these tables that Kepler used to put the Copernican revolution on a firm footing.

Kepler found it relatively easy to reconcile the new heliocentric system with the planetary motions recorded by Tycho — except for those of Mars. It was Kepler's torturous work on Mars that led him to his great discovery:

The planets move in ellipses, not circles; and their orbital speeds change predictably as their distances from the sun vary. These insights allowed the most accurate computations of celestial positions — and did it in the Copernican framework.

On the heels of Kepler's success a new tool, the telescope, became available to astronomers. Although the first telescopes were too crude to permit interesting discoveries on Mars, it wasn't long before improved instruments revealed a planet much more interesting than the bland Venus. By 1677 astronomers had discovered polar ice caps, along with bright and dark markings that were regarded as deserts and seas. These features allowed astronomers to measure the length of the martian day, which at 24 hours, 37 minutes is remarkably similar to Earth's. The ice caps were seen to grow and shrink with the martian seasons, and the bright and dark patches also varied from year to year. These characteristics suggest a striking resemblance to Earth. That resemblance was briefly enhanced by the "discovery" of another type of feature — the canals of Mars.

The story of the canals is among the most bizarre episodes in the annals of science. The term "canals" first appeared in work of an Italian astronomer, Pietro Secchi, during the 1860s. He noted faint linear features on the planet, which he called *canali* (channels). The term was picked up by the great astronomer Giovanni Schiaparelli, who noted a number of such features. As word of Schiaparelli's observations spread, other astronomers strained to see the canals, which seemed to appear only under the very best observing conditions. One astronomer who heard of the canals, and subsequently believed he saw them, was Percival Lowell. Lowell soon became convinced that the canals were created by an advanced race of Martians to irrigate the planet's vast deserts. The wealthy Bostonian built an excellent observatory in Flagstaff, Arizona, to acquire high-quality observations of the planet. He wrote three popular books expounding his theory of a martian civilization. Many experienced astronomers were skeptical of the existence of the canals — to say nothing of a martian civilization — and today they are regarded as a product of optical effects and wishful thinking. (Another of Lowell's theories fared much better. He predicted the existence of a planet beyond Neptune, which he called Planet X. In 1930, fourteen years after his death, an astronomer named Clyde Tombaugh discovered Pluto from the observatory Lowell founded.) Despite the skepticism of many, however, the idea of life on Mars became ingrained in popular, and even scientific, notions about the planet. In fact, even the existence of canals had some proponents until they were definitely ruled out by pictures returned by the spacecraft that began visiting Mars in the 1960s.

To date, thirteen spacecraft have been sent to Mars; at this writing a fourteenth, *Mars Observer*, is scheduled to arrive in 1993. The first few each made a single pass by the planet, returning images of a landscape that, somewhat disappointingly, resembled the moon. Not only did these craft fail to find canals, or any other signs of life, they detected a thin atmosphere of carbon dioxide. With sub-zero temperatures, the planet resembled a moonlike frozen desert. It all added up to an environment quite inhospitable to any form of life, to say nothing of advanced civilizations. (Table 4-2 lists important facts about Mars.)

But in 1971 the *Mariner 9* orbiter made discoveries that struck the astronomical community like a thunderclap. The first few weeks of the mission were inconclusive, as the probe arrived during a dust storm that enveloped the entire planet, obscuring the cameras' view. As the dust cleared, four dark spots appeared in the northern hemisphere. It soon became apparent that these spots were enormous mountains — ancient volcanoes — that reach up to 17 miles (27 kilometers) above the surface. *Mariner 9* also found an enormous crack on the surface, dubbed Valles Marineris, which, if transported to Earth, would run the length of the continental United States. But the most important discovery was a wide variety of features that could only be formed by running water: flood plains, canyons, and dried-up river beds.

The unmistakable signature of flowing water allowed scientists to speculate that rudimentary life forms may have survived the planet's transition from an ancient tropical paradise to its present apparently barren state. In 1975 the most ambitious planetary mission ever undertaken sent two pairs of Viking probes to the planet. Each Viking consisted of an orbiter and a sophisticated lander. The landers served as weather stations, biological laboratories, and even geological probes. The biological experiments gave somewhat confusing results; today, the consensus is that they did not detect signs of life. The two Viking missions were designed to last a minimum of ninety days, but they far exceeded this specification, sending us several years' worth of information on weather, the martian soil, atmospheric composition, and geology. So why is Mars red? Evidence from the Viking landers indicates that iron-rich clays cover much of the surface, giving Mars its rusty hue.

Enjoying the Martian Sky Show

The motions of Mars, which mystified all astronomers prior to Kepler, continue to give us a delightful exhibition of celestial geometry. Let's look at a typical apparition of Mars, that of 1994-95, beginning with its first appearance in the east at dawn.

As we saw in the introduction, Earth's orbital motion makes the stars appear noticeably farther west each week. This westward motion tends to carry the planets along,

Table 4-2. Facts About Mars

Diameter:	6,786 kilometers 4,217 miles 53% that of Earth
Surface temperature:	-55° C -67° F
Surface atmospheric pressure:	¹⁄₁₅₀ that of Earth
Atmospheric composition:	95% carbon dioxide 2.7% nitrogen
Moons:	Phobos Deimos
Rotation period:	24.62 hours
Sidereal orbital period:	686.98 days or 1.88 years
Synodic orbital period:	779.94 days or 2.14 years Longest of any planet
Mean distance from sun:	227.9 million kilometers 141.6 million miles 1.52 times that of Earth
Orbit inclined to Earth's:	1.85 degrees

too — except in the case of Mars, whose eastward motion now counteracts the effect of Earth's travels. As Figure 4-2 shows, Mars lies just above the eastern horizon, below the constellation Pegasus, in early June. By September, when Pegasus has moved to the evening sky, Mars can be found in Gemini — and is still visible only in the morning sky. By December, however, Mars begins to slow down and is carried westward with the stars; the planet also begins to brighten noticeably. In November, it lies in the constellation Leo, having cruised through about one-third of the sky since June. Now it is slowing down, each week appearing higher in the sky with Leo. On January 10, 1995, Mars can be seen in the east by about 10 P.M.; this is a good time to observe the highlight of the planet's show. Its eastward motion grinds to a halt shortly after this date. This stationary point is where the motion reverses and the retrograde loop begins. On February 12, Mars is at opposition — opposite the sun, rising at sunset. Mars continues westward, but slows down and finally halts at the end of March. It now accelerates eastward, once again defeating the general westward drift of the stars. The Red Planet lingers in the western sky until it's lost in the twilight glow around the end of the year.

Let's see how the insight of Aristarchos and Copernicus clarifies what Mars has done. Again, we'll look at the 1995 opposition as it appears from space. Figure 4-3 illustrates the basic geometry that accounts for the Red Planet's changes in brightness and motion. In June, 1994, Mars is over half an Earth orbit away from us. Though slower than Earth, it has sufficient orbital velocity to maintain a good distance from us. As the two curved paths become nearly parallel, Mars begins to slow down against the background stars. For a few days in January, the planets seem to pace each other, and Mars grinds to a halt; this is the beginning of the retrograde loop, shown in detail at the top of Fig. 4-3. By February Earth's greater speed has minimized the remaining gap between the planets, and naturally, the reduced distance results in Mars brightening. The apparent reversal of Mars's motion is, of course, the result of our planet overtaking it. (You can see a similar effect when you pass a car on the freeway, but please, let someone else do the driving!) Eventually, Earth "turns the corner" in its orbit and the Red Planet's apparent motion again tracks its orbital motion. Note the position of Mars on February 12 in Figure 4-4. This is opposition, when Mars lies opposite the sun in Earth's sky. Since our

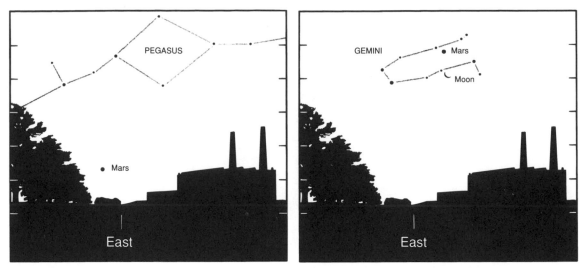

Figure 4-2. *Left: In June of 1994, Mars (+1.2) barely makes it above the horizon thirty minutes before sunrise. It lies below the prominent star pattern known as the Great Square of Pegasus. Right: Three months later Mars hasn't brightened a bit and doesn't appear much higher in the sky, but it has moved on to the constellation Gemini.*

distance is at its minimum then, that's when Mars is at its brightest.

You may have noticed that in Table 4-1 we gave the opposition distance. We wouldn't have bothered if that distance was always the same, right? Remember that Mars does not have a circular orbit. Figure 4-4 shows an interesting consequence of the Red Planet's lopsided orbit: Not all oppositions are alike. Mars's distance to the sun varies considerably, from a perihelion (closest approach) distance of 128,280,000 miles (206,445,000 km) to an aphelion 155,524,000 miles (249,828,000 km). If Mars reaches opposition when it is near perihelion, it is nearest both the sun and us and appears brightest. Unfortunately, the oppositions of 1993, 1995, and 1997 occur when Mars is near aphelion, the point in its orbit when it is *farthest* from the sun. This has a big effect on the brightness (and, with a telescope, apparent size) of Mars. After 1997, the oppositions improve as they occur progressively nearer perihelion — something to look forward to. Wait till you see how bright Mars gets in 2003!

Color Plate 1. *An aurora splashes color through the sky above Wasilla, Alaska. Solar flares and other activity on the sun cause streams of electrons to cascade into the atmosphere, setting air molecules aglow. Photo by Forrest Baldwin.*

Color Plate 2. *Sunlight refracted and reddened by passing through earth's atmosphere tints the moon during the total lunar eclipse of July 1982. The shading is lighter where the moon lies closer to the edge of the earth's umbra, the darkest portion of its shadow. Photo by Fred Espenak.*

Color Plate 3. *The "diamond-ring effect," a small bright point of light from the emerging sun, marks the end of totality during the solar eclipse of July 1991. The pale glow of the inner corona and flame-like prominences — huge arcs of gas suspended above the sun's surface — are also visible. Photo by Fred Espenak.*

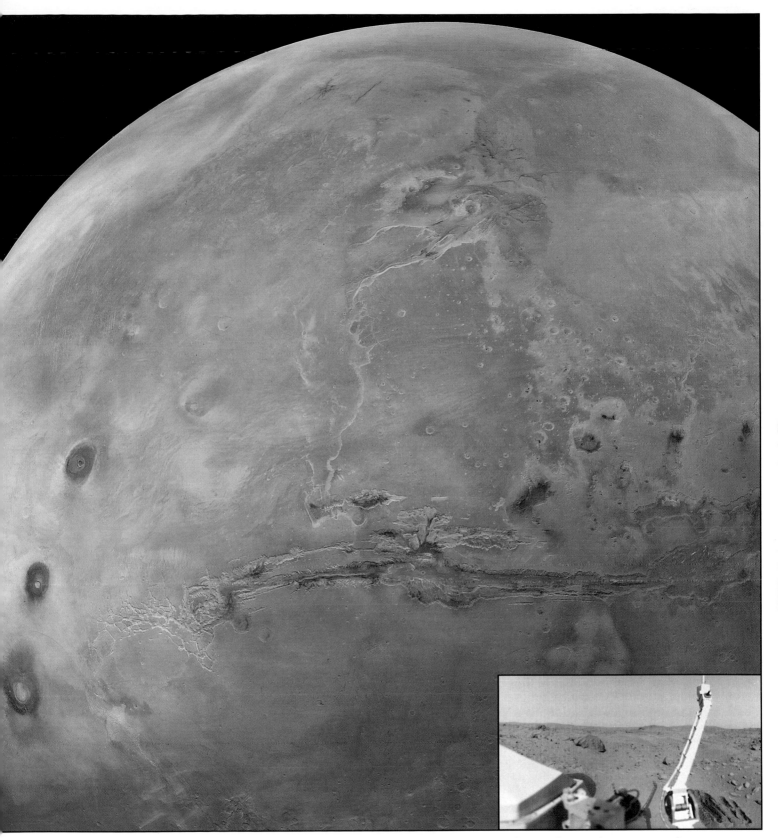

Color Plate 4. *The huge Valles Marineris canyon system runs across the center of this mosaic of 102 Viking Orbiter images of Mars. White wisps of cloud surround three enormous volcanoes at lower left, each 16 miles (25 kilometers) high. On earth, this chain of volcanoes would stretch from Maine to South Carolina. Inset: The surface of Mars as viewed from the Viking 1 lander in 1976. Trenches dug by a robot arm allowed the craft to analyze the soil's composition and search for microbes. Boulders and sand dunes dominate the landscape. Fine dust lofted by winds gives the sky its alien hue. Photo by Alfred McEwen, USGS; Inset Photo by NASA/JPL.*

Color Plate 5. *The Voyager spacecraft returned unprecedented views of Jupiter, Saturn, Uranus, and Neptune between 1979 and 1989. This view of Jupiter's swirling, colorful atmosphere was returned by Voyager 1 in 1979, still 20 million miles from the giant planet. The large oval-shaped feature is the Great Red Spot, a giant storm larger than earth and observed for centuries. Photo by NASA/JPL*

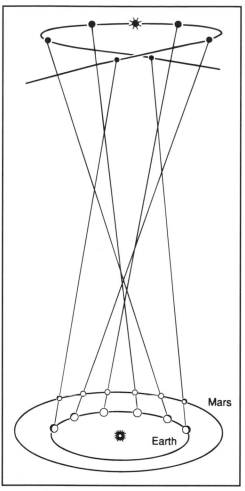

Figure 4-3. *The remarkable loop of Mars is less mystifying when seen from space. The positions of the planets in this cartoon reflect those of the 1995 opposition; the loop at the top shows both the path Mars takes through the stars and the planet's changes in brightness. As Earth catches up in January 1995, Mars appears to halt its motion among the stars. As we overtake it, its motion reverses. By April our orbit carries Earth "around the corner," and the Red Planet's orbital motion again directs its apparent motion.*

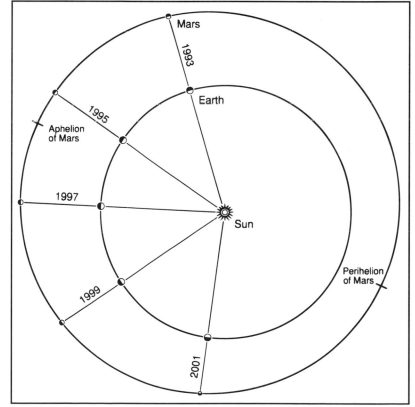

Figure 4-4. *The orbits of Earth and Mars are not concentric circles, so the distance between the two planets varies from opposition to opposition. As shown here, the distance between Earth and Mars decreases from the 1997 opposition on. The orbits of Mercury and Venus are omitted for clarity.*

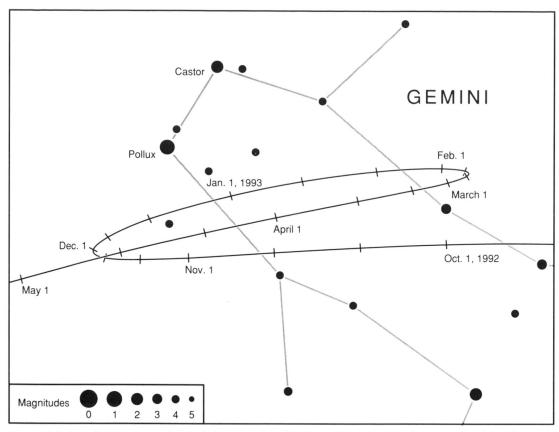

Figure 4-5. *Our first two oppositions of Mars come during the cold winter months, when the planet lies in two prominent constellations. This will make the retrograde loops especially easy to follow. Top: Mars takes a whirl through Gemini in 1993. Between September and May Mars is noticeably brighter than any star shown here, and in January it outshines the brightest stars in the sky. Bottom: The 1995 apparition is the worst of the decade. Still, Mars outshines the bright star Regulus throughout its loop in Leo.*

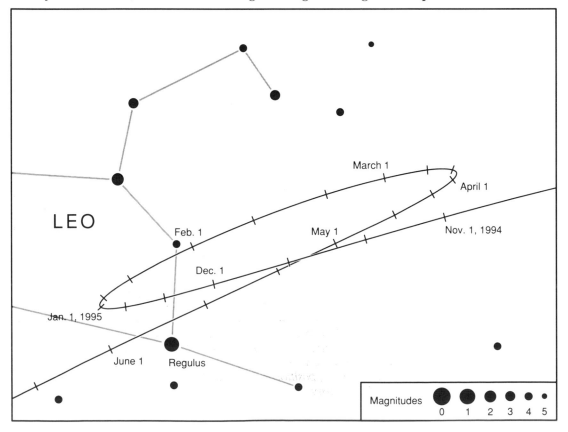

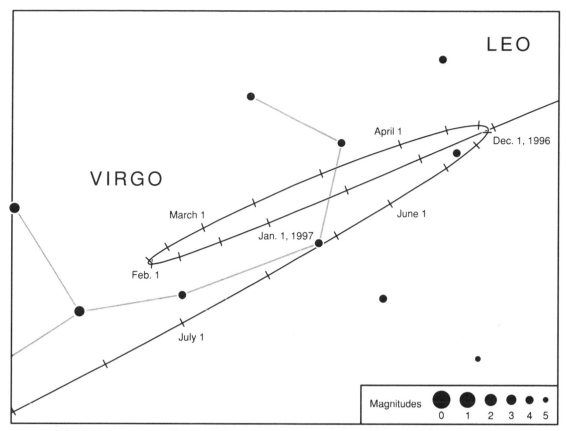

Figure 4-6. *The oppositions of 1997 (top) and 1999 (bottom) are both in the sprawling constellation Virgo, one of the longer constellations of the zodiac. The bright star Spica serves as a handy gauge of the planet's motion, especially when Mars swings close to it in 1999.*

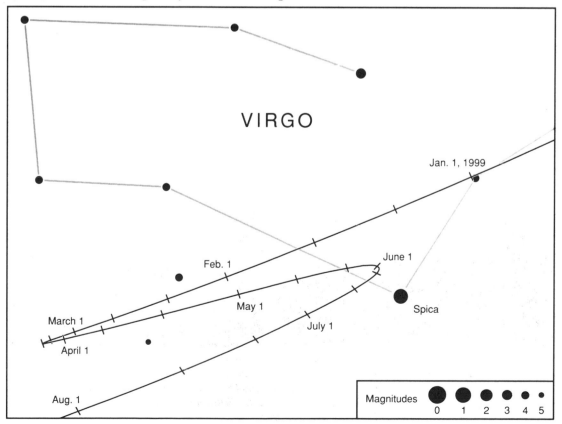

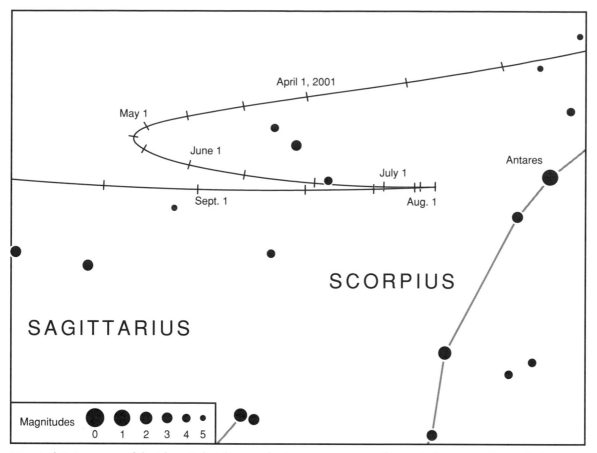

Figure 4-7. *In terms of the planet's brightness, the 2001 opposition of Mars is the best in the period covered by this book. However, because the planet now courses through the southernmost constellations, it does not rise as high above the horizon as it did in earlier years. Look for the star Antares, the brightest star in the constellation Scorpius. Its name, which means "rival of Mars," invites comparison between these two fiery objects.*

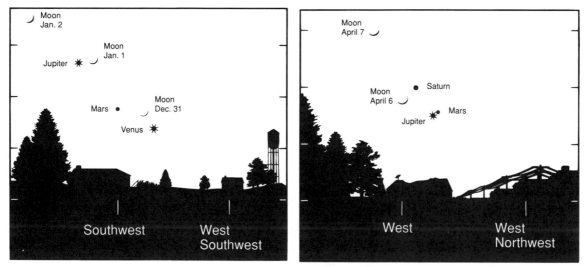

Figure 4-8. *As Mars lingers in the western twilight before making its final plunge toward the sun, it's joined by the moon and other planets. Two of the best gatherings are shown here. Left: The moon, brilliant Venus (-4.5), and bright Jupiter (-2.1) join Mars (+1.2) about thirty minutes after sunset during the last days of 1997. Right: The moon joins Jupiter (-2.0), Saturn (+0.3), and Mars (+1.4) as seen thirty minutes after sunset on April 6, 2000. Binoculars help.*

5 DISTANT GIANTS: JUPITER AND SATURN

After following the breathless wanderings of Mars, we turn now to planets that ply the sky at a much more leisurely pace. Jupiter and Saturn, the largest planets of the solar system, lie much farther from the sun — and us — than Mars. Jupiter's lane of the solar system racetrack is about five times the size of Earth's orbit, and Saturn's track is twice the diameter of Jupiter's! As a result both of their longer circuits and slower speeds, the wanderings of Jupiter and Saturn through the constellations are much less dramatic than the splendid whirl of Mars. As outer planets, they exhibit most of the same motions as Mars: they go through retrograde loops each year as Earth laps them on the solar system racetrack and, like Mars, are brightest near opposition. But unlike Mars, which courses through several constellations during each apparition, these planetary giants make seasonal appearances that approximate those of the background stars. For example, Jupiter will travel through roughly one constellation each year, Saturn about half that. So observers can count on seeing Jupiter and Saturn only slightly later each successive year.

Besides their stately motions, Jupiter and Saturn make a more significant departure from the planets we've discussed up to this point. They are the nearest of the four *gas giants* of the outer solar system, huge planets largely composed of hydrogen and helium, the two lightest elements in the cosmos. They stand in sharp contrast to the small, dense, rocky worlds that orbit much closer to the sun. And they are enormous: Jupiter, with a diameter eleven times the size of Earth's, contains more than twice as much matter as all the other planets put together. It holds 318 times the mass of our planet in a volume large enough to contain 1,300 Earths. On the basis of mass alone, one could argue that the planetary system consists of Jupiter plus debris! Saturn is a distant second, weighing

in at 95 Earth-masses with a volume large enough to hold 750 earths.

During the period covered in this book, Jupiter and Saturn have a special relationship, as shown in Fig. 5-1. From now until the year 2000 they appear closer and closer together as Jupiter gradually catches up to a slow-moving Saturn. Jupiter begins 1993 in the spring constellation Virgo; by 2001, it has passed Saturn and lies in Gemini, the frigid twins of the winter sky. Saturn, on the other hand, drifts along at about one-third Jupiter's pace. It lies among the stars of the faint autumn constellation Aquarius in January 1993, and eight years later has traveled only as far as Taurus, the celestial bull of late autumn. But, as we'll see, Saturn has a few interesting tricks to keep us entertained along the way.

Tables 5-1 and 5-2 list the dates when the two planets are at opposition, directly opposite the sun and visible to the south at midnight. (Keep in mind that each month after those dates, the planets can be found due south two hours earlier.) The horizon scenes in Figs. 5-2 through 5-5 depict some of the best arrangements of Jupiter and Saturn with other planets, particularly Venus, and also illustrate how the pair gradually closes the gap between them. Also, the yearly summaries in the Appendix will give you a good idea of when and where you'll find these giant planets. For the next few years they'll be easy to spot since they're wandering among faint constellations whose stars they far outshine, a fact that should enable you to quickly locate them.

Celestial Kingpins

It is somewhat curious that the most distant planets known to the ancients were often associated with the most powerful members of ancient pantheons. Jupiter in

Table 5-1. Oppositions of Jupiter, 1993–2001

Opposition date	Constellation	Mag.	Nearest to Earth	Distance from Earth when closest
Mar. 30, 1993	Virgo	-2.5	Mar. 31	666 million kilometers 414 million miles
Apr. 30, 1994	Libra	-2.5	May 1	662 million kilometers 411 million miles
June 1, 1995	Scorpius	-2.6	June 2	647 million kilometers 402 million miles
July 4, 1996	Sagittarius	-2.7	July 5	626 million kilometers 400 million miles
Aug. 9, 1997	Capricornus	-2.8	Aug. 10	606 million kilometers 377 million miles
Sept. 16, 1998	Aquarius	-2.9	Sept. 15	593 million kilometers 368 million miles
Oct. 23, 1999	Pisces	-2.9	Oct. 22	593 million kilometers 368 million miles
Nov. 28, 2000	Taurus	-2.9	Nov. 26	606 million kilometers 377 million miles

particular was accorded the highest place in the pantheons of several civilizations. Their brilliance can't be the reason since the sun, moon, and even Venus far outshine them. The regal pace with which they drift through the constellations of the zodiac is a more likely explanation.

To the ancient Chinese Jupiter was *Sui hsing*, the "Year Star." As the name suggests, Jupiter served a calendrical function from about the seventh century B.C. based on its twelve-year cycle of residence in certain zones of the sky. These regions followed the celestial equator, an imaginary line midway between the north and south celestial poles, as opposed to the ecliptic. Jupiter comes to opposition twelve times in just over twelve years, and the twelfth opposition returns the planet to nearly the same stars as the first. The Chinese came to associate each of the twelve sky intervals with an animal, and as Jupiter moved from one zone to another the year took on the characteristics of whatever animal realm Jupiter was in. So the traditional Oriental cycle of names for the year — Year of the Rat, Year of the Ox, etc. — derives in part from Jupiter's twelve-year journey through the stars.

For the ancient Egyptians, Jupiter was yet another manifestation of the sky god Horus, whom they saw in Mars and Saturn as well. Jupiter was "Horus Who Bounds the Two Lands," "Horus Mystery of the Two Lands," and "Horus who Illuminates the Two Lands." The "two lands" were Upper and Lower Egypt, and all of these names date from the New Kingdom (between 1580 and 1085 B.C., comprising the eighteenth to twentieth dynasties). Another, perhaps older, name calls Jupiter "Star of the Southern Sky."

A more striking example of Jupiter's dominance is in the story of Babylonian god Marduk, whom we met briefly in Chapter 1. Marduk was regarded as the creator of the world, who established cyclic order in the cosmos by vanquishing Tiamat, one of two primordial creative elements whose blind chaotic energy dominated the early universe. In reward for this deed he was made lord and ruler of the gods. Marduk adopted Jupiter (*Niburu*) as his personal star, a "night watchman" whose steady course through the sky guided the stars and planets and preserved order. In twelve years Jupiter completes the circuit through the stars the sun makes every twelve months, significant because of the importance of the number twelve in Babylonian culture. And of all the planets, Jupiter most closely follows the sun's path through the stars.

The Greeks, too, connected Jupiter with the king of the gods. The Greek name Zeus is derived from the

Table 5-2. Oppositions of Saturn, 1993-2001

Opposition date	Constellation	Mag.	Nearest to Earth	Distance from Earth when closest
Aug. 19, 1993	Capricornus	+0.3	Aug. 20	1,316 million kilometers 818 million miles
Sept. 1, 1994	Aquarius	+0.5	Sept. 1	1,302 million kilometers 809 million miles
Sept. 14, 1995	Aquarius	+0.7	Sept. 14	1,287 million kilometers 800 million miles
Sept. 26, 1996	Pisces	+0.5	Sept. 26	1,271 million kilometers 790 million miles
Oct. 10, 1997	Pisces	+0.2	Oct. 10	1,255 million kilometers 780 million miles
Oct 23, 1998	Pisces	0.0	Oct. 23	1,241 million kilometers 771 million miles
Nov. 6, 1999	Aries	-0.2	Nov. 6	1,228 million kilometers 763 million miles
Nov. 19, 2000	Taurus	-0.4	Nov. 19	1,217 million kilometers 756 million miles
Dec. 3, 2001	Taurus	-0.4	Dec. 3	1,209 million kilometers 751 million miles

Sanskrit *Dyaus*, "shining one." (Our name for the planet stems from the Sanskrit *Dyauspitar*, through *Zeuspater*, "father Zeus.") Zeus was actually a third-generation god in Greek mythology. In one Greek creation myth, Gaia, Mother Earth, arose from chaos and gave birth to Uranus. Together Gaia and Uranus engendered such strange creatures as the one-eyed Cyclopes before giving birth to the Titans. But Uranus exiled the Cyclopes, and in anger Gaia persuaded the Titans to attack Uranus. Kronos, the youngest Titan, led the rebellion and became the leader. But the dying Uranus had prophesied that Kronos (equivalent to the Roman Saturn) would be similarly overcome by his offspring. Kronos resolved to swallow his children to avoid that fate, but his wife Rhea hid Zeus, their youngest. Upon maturity, Zeus returned to confront his father. Rhea gave Kronos a potion that made him vomit up Zeus' siblings, who asked him to lead them in a rebellion. After a long (you might say titanic) struggle, the children of Kronos overthrew the Titans, and Zeus assumed rule over the universe. Zeus was thought to make his presence known to mortals with his lightning bolts; the Norse god of thunder, Thor, gives his name to the fifth day of the week.

To the ancient Chinese, Saturn was the *Chin hsing*, "quelling star" or the "yellow oppressor"; despite these names, Saturn was a minor object for the astrologically-minded Chinese. Saturn was for Egyptians another manifestation of Horus — "Horus the Bull of the Sky." Saturn was known to the Babylonians as *Ninib* or *Ninurta*, originally a war-god as well as the god of fertility and agriculture, the south wind personified. As mentioned above, the Greeks connected the planet with Kronos, the father of Zeus. However, when transported to the Italian peninsula (where the planet got the name we now use), Saturn seems to have become associated with agriculture as well, perhaps by blending with an Etruscan god. The Roman Saturnalia festival, a sort of harvest celebration held just prior to the winter solstice, was celebrated with unusual abandon — even slaves were given a day of freedom. Today, many of us may have similar feelings of release on the last day of the week, which gets its name from the god of the Saturnalia.

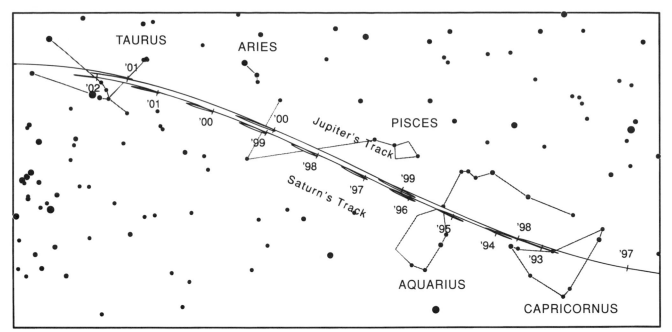

Figure 5-1. *Top: Jupiter, a member of the "Constellation-a-Year Club," drifts through two-thirds of the sky and visits eight of the twelve constellations of the zodiac from 1993 through 2001. Saturn, a much more leisurely wanderer, travels less than half that distance. Tick marks indicate positions at the beginning of each year. This illustration shows Saturn's entire course during the period and much of Jupiter's track. Notice the retrograde loops and how the two planets meet in Taurus toward year's end in 2000. Bottom: The relative positions of Jupiter and Saturn for each of their forthcoming oppositions. Jupiter gains a little on Saturn each year and finally laps Saturn in the year 2000.*

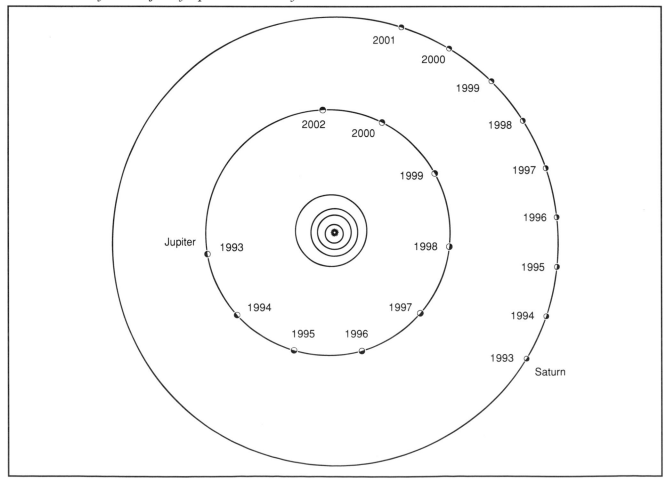

There are few indications that Jupiter or Saturn played any roles in the astronomy, mythology, or astrology of the Americas. Perhaps their relatively tranquil motions failed to incite speculation about their divine qualities.

King of the Planets

As we saw in the previous chapter, efforts to explain the motions of the planets ultimately gave humanity (at least the European portion) a new perspective on its place in the universe. But Mars played a much bigger role in the Copernican revolution than either Saturn or Jupiter. It wasn't until the invention of the telescope that these distant, giant planets came into their own scientifically.

In the spring of 1609 word of a new invention reached Galileo Galilei, the Professor of Mathematics at the University of Padua in Italy. He fashioned his first crude telescope from glass lenses and lead pipe, but quickly improved on this first model. Galileo didn't wait long to turn his new telescope toward the heavens. His description of his first Jupiter observations remains compelling over three hundred seventy years later:

On the 7th day of January in the present year [1610] . . . the planet Jupiter presented itself to my view and, as I had prepared for myself a very excellent instrument, I noticed a circumstance which I had never been able to notice before, namely, that three little stars, small but very bright, were near the planet; and, although I believed them to belong to the number of the fixed stars, yet they made me somewhat wonder, because they seemed to be arranged exactly in a straight line . .When on January 8, led by some fatality, I turned again to look at the same part of the heavens, I found a very different state of things, for there were little stars all west of Jupiter, and nearer together than on the previous night. . .

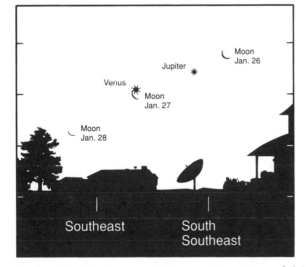

Figure 5-2. *Top: The waning moon joins Venus (-4.3) and Jupiter (-1.9) in the eastern sky forty-five minutes before dawn on January 27, 1995. Left: On October 26, 1995, look for Jupiter (-1.9) above the young moon, low in the southwest thirty minutes after sunset. Faint Mars (+1.4) lies below it. Right: At the same time and date, turn east to find Saturn (+0.8) rising beneath the constellation Pegasus.*

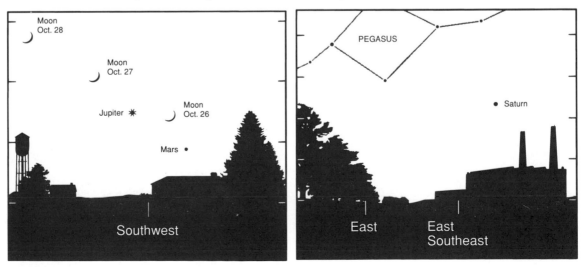

Galileo was quick to realize that he was seeing orbital motion edge-on, and recognized that Jupiter and its four satellites were a perfect analog of the Copernican model of the solar system. In addition, they were an unequivocal demonstration of celestial motion that did not have the earth at its center. Galileo called the four bright moons the "Medicean stars" after his patrons, the Medici, but today these satellites — Jupiter's largest — are called the Galilean moons in honor of their discoverer.

Better telescopes soon revealed fascinating details on the planet. The first features noted were the bands that are easily seen with the small telescopes that are common in hobby shops today; later, those simple bands were resolved into swirls and festoons. The most remarkable feature of the planet, the Great Red Spot, may have been noticed by Robert Hooke in 1664; it was certainly seen the following year by Giovanni Cassini. This amazing feature,

several times larger than our planet, is akin to an enormous hurricane that has raged on the planet, presumably, for over three hundred years. Features like the Great Red Spot allowed astronomers to time the rotation of the planet to an amazingly rapid 9.8 hours. Jupiter's rapid spin causes the fluid planet to bulge noticeably at the equator. The planet does not rotate as the solid inner planets do; each band represents a slightly different rotation rate.

The first telescopic observations of Saturn were also made by Galileo. Because of the relatively poor quality of his telescope, he didn't have a clear view of the now-famous rings. He did detect two points of light on either side of the planet, but took them to be moons. He was mystified when, two years later, those satellites had completely disappeared from his view. He recovered sight of them in 1616, but was puzzled that they didn't drift back and forth as did Jupiter's satellites. For some time, Saturn

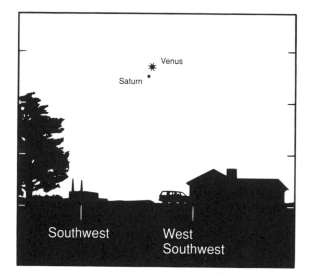

Figure 5-3. *Top: Venus (-4.1) and faint Saturn (+1.2) form a close evening pair on February 3, 1996. Look for them in the southwest in the hour after sunset. The moon, one day from full, is low in the east behind you. Left: Venus (-4.2) and Jupiter (-2.1) are very close on the morning of April 23, 1998. The brilliant pair hovers near the waning moon; look thirty minutes before dawn. Right: Just over a month later, on May 29, 1998, Venus (-4.0) and Saturn (+0.6) are very close in the morning sky.*

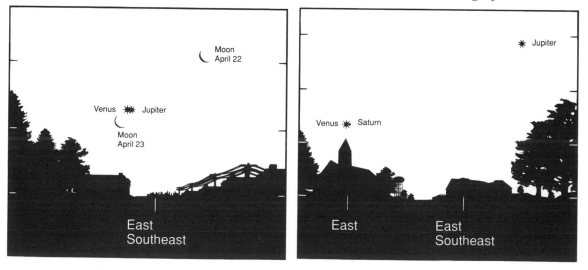

Table 5-3. Facts About Jupiter and Saturn		
	Jupiter	**Saturn**
Diameter:	142,984 kilometers 88,850 miles 11.21 times that of Earth	120,536 kilometers 74,901 miles 9.45 times that of Earth
Atmospheric composition:	90% hydrogen 10% helium	97% hydrogen 3% helium
Number of moons:	16	18
Largest moons:	Ganymede Callisto Io Europa	Titan
Rotation period:	9.84 hours	10.23 hours
Sidereal orbital period:	4,332.71 days or 11.86 years	10,759.5 days or 29.46 years
Synodic orbital period:	398.88 days or 1.09 years	378.09 days or 1.04 years
Mean distance from sun:	778.33 million kilometers 483.63 million miles 5.2 times that of Earth	1,426.98 million kilometers 886.72 million miles 9.54 times that of Earth
Orbit inclined to Earth's:	1.31°	2.49°

was thought of as a "triplet" planet. In 1656, when the Dutch scientist Christiaan Huygens viewed the planet, the rings were not visible but a thin dark shadow was. Huygens realized that he was seeing a ring system edge on. As the planet continued in its orbit, the terrestrial viewing angle gradually changed and the rings reappeared. Huygens published his findings in 1659, but for a while there was resistance to the idea of a system of rings that were unconnected to the planet.

Improved observations settled the matter, and by 1850 earth-bound observers had detected three separate components to the rings. About that same time, astronomers came to understand that the rings were actually billions of individual particles orbiting independently. Although the existence of Saturn's ring system is familiar to us all, they are truly wonderful to contemplate. The main rings cover an area of just over fifteen billion square miles (forty billion square kilometers), or eighty times the total surface area of the Earth. The rings span 171,000 miles (274,000 kilometers) — about seventy percent of the distance separating Earth and the moon — but are thought to be just 33 feet (10 meters) thick. If you made a scale model of the ring system from material as thick as a quarter, the model would have to be ten miles (sixteen

kilometers) across. Even if you used material as flimsy as plastic food wrap, a properly scaled model would be nearly a quarter of a mile (a third of a kilometer) across. Little wonder the rings vanish when they present themselves edge-on to Earth.

Telescopic observations, along with the laws of physics derived by Kepler and Newton, enabled earth-bound astronomers to discern the gross properties of the giant planets — size, mass, and composition — which are summarized in Table 5-3. With the advent of the Space Age, astronomers no longer had to be content to observe the planets from afar. Spacecraft were sent to take closeup pictures of both planets and their moons, and also to directly measure their environments with scientific instruments. The first such probes were the *Pioneer 10* and *Pioneer 11* spacecraft. Launched in 1972 and 1973 respectively, these robotic investigators carried eleven instruments, from cameras to particle detectors. Owing to the gas planets' large distances from the sun, the probes had to rely on nuclear power; the solar energy at Jupiter is only 4 percent the amount available at Earth. *Pioneer 10* reached Jupiter after a twenty-one-month trip that included a proof-of-concept passage through the asteroid belt. One major discovery was the intensity of the radiation belts created by Jupiter's powerful magnetic field; a significant amount of scientific data was lost when the radiation disrupted spacecraft operation. *Pioneer 10* was followed a year later by *Pioneer 11*, which continued on to Saturn. Although the Pioneer missions made major scientific contributions in their own rights, their main contribution was blazing a trail for the spectacularly successful Voyager missions that followed.

By practically any measure, the Voyager Project must be considered the greatest scientific undertaking in the history of mankind. For a modest per capita annual

investment — roughly half the cost of a candy bar — the United States sent two probes to four planets, discovered dozens of moons, acquired awe-inspiring pictures, and collected a truly mind-boggling wealth of data from which scientists will be teasing new discoveries well into the next century. In fact, launched in 1977, the two Voyagers continue to collect data about the solar environment and, with luck, will serve as humanity's first inter*stellar* probes early next century!

NASA scientists are fond of saying that the study of other planets illuminates our own; for instance, the global

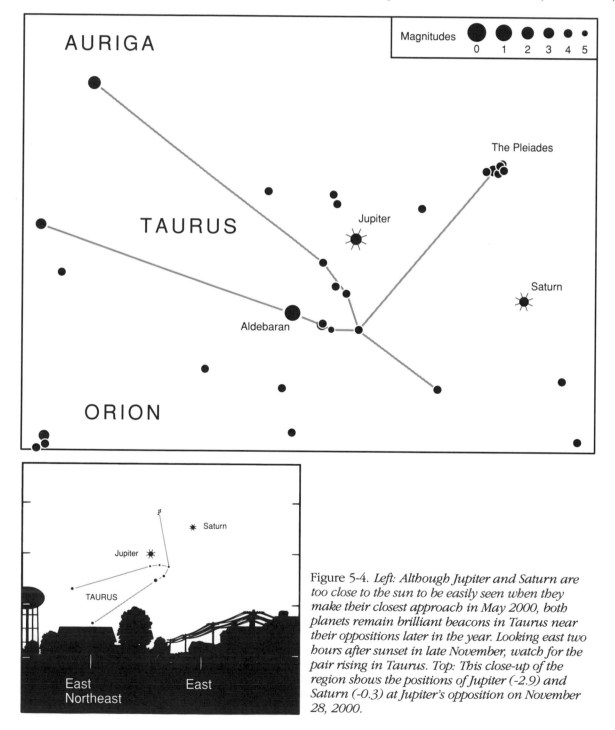

Figure 5-4. *Left: Although Jupiter and Saturn are too close to the sun to be easily seen when they make their closest approach in May 2000, both planets remain brilliant beacons in Taurus near their oppositions later in the year. Looking east two hours after sunset in late November, watch for the pair rising in Taurus. Top: This close-up of the region shows the positions of Jupiter (-2.9) and Saturn (-0.3) at Jupiter's opposition on November 28, 2000.*

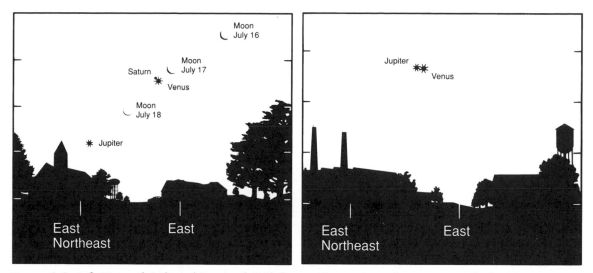

Figure 5-5. *Left: Venus (-4.1) and Saturn (+0.2) form a close pair on the morning of July 15, 2001, thirty minutes before sunrise. Right: Venus (-4.0) and Jupiter (-2.0) are closest on the morning of August 5, 2001. Saturn (+0.1) lies above the pair, and the waning moon joins the trio on the 14th. Look in the east in the hour before dawn.*

perspective on the weather of Jupiter and Saturn has certainly refreshed and enhanced our understanding of Earth's climate. But the exploration of Jupiter and Saturn also broadened our perspective on the laws of nature by showing us the behavior of matter under conditions that cannot be reproduced on Earth. An excellent example was the discovery of active volcanoes on Io, one of Jupiter's Galilean moons. At least two types of volcanoes have been found, and while one type resembles terrestrial volcanoes, neither is actually produced by the same internal engines as Earth's eruptions. Similarly, the dazzling closeups of Saturn's intricate ring system forced physicists to push their dynamical theories to new limits to accommodate the things the Voyager cameras showed them (Color Plate 6). Spacecraft cameras resolved each of the larger rings into thousands of tiny ringlets, some kinked and clumpy, others distorted by the presence of moonlets within the rings. Waves and corrugations moved throughout the rings.

One of the most ingenious methods of investigation was to use minute changes in the spacecraft's radio signals to measure the structure of the planets' gravitational fields. This in turn enabled scientists to describe in amazing the detail the structures of the planets themselves. Both may be compared to plums, with rocky cores being the pits, and the lighter elements helium and hydrogen being the analogs of the sweet meat of the fruit. (The portions visible to our immediate inspection resemble the skin of the fruit.) While hydrogen and helium may suggest buoyant children's balloons, these elements are in a gaseous form in only the outermost reaches of the gas giants. A thousand kilometers or so beneath the cloudtops, these elements are compressed to a liquid state by the tremendous weight of the atmosphere above. Even deeper, the

hydrogen is squeezed into a peculiar state called liquid metallic hydrogen, which conducts electricity much as a metal would. Beneath this layer is a rocky core larger than our planet and several times its mass.

Dance of the Giants

As we said earlier in the chapter, Jupiter and Saturn are much more sedate than Mars in their travels through the sky. This makes it easier for even casual stargazers to become acquainted with them and anticipate their annual appearances.

There are twelve constellations in the zodiac, and Jupiter completes an orbit every twelve years, so Jupiter's travels can be summed up in the rule of thumb, "a constellation a year." Jupiter lies in the spring constellation Virgo in 1993, and so appears in the east after sunset in late April and early May. Each successive year, it makes its evening appearance about a month later. By the year 2000, when you have to wait until November for an evening glimpse of the planet, it will have caught up to Saturn. During the period covered in this book, it will take you through two-thirds of the constellations of the zodiac; odds are, it'll pass through your astrological "birth sign." (If not, Chapter 6 will guide you to your astrological constellation — and discuss why it's not really your zodiacal sign.)

Saturn, on the other hand, is a fall object for the balance of the century. It starts 1993 in the constellation Aquarius, and can be seen in the evening sky by mid-October each year. By 1998, when Saturn lies in eastern Pisces, Jupiter lies in western Pisces, and you'll be able to watch the slow chase in the evening sky until Jupiter catches up in the fall of 2000, (Figs. 5-2 through 5-5).

The grand conjunction of Jupiter and Saturn that occurs in 2000 doesn't happen very often, and it's easy to see why. By dividing Saturn's orbital period of twenty-nine years into 360 degrees, you'll find that Saturn moves eastward through the stars by about twelve degrees each year. Do the same for Jupiter and the result is an annual eastward motion of about thirty degrees per year. So Jupiter gains about eighteen degrees on Saturn annually, which means that about twenty years must pass before it will pass Saturn again.

While you're waiting for that splendid conjunction there's another interesting phenomenon to take note of, one created by Saturn's icy rings. Although you need a telescope to actually observe the rings, you can appreciate the way they affect the planet's brightness. As we discussed earlier, twice during Saturn's twenty-nine-year orbit its rings appear edge-on from Earth. Saturn spins on an axis tilted about twenty-seven degrees to its orbit, and this tilt gives us different perspectives on the rings system as Saturn circles the sun. At Saturn's equinoxes, when neither its north or south pole angles into the sun, we pass through the plane of the rings and they appear edge-on. Because the rings are made of bright, icy material, they reflect sunlight well and contribute significantly to Saturn's overall brightness when the angle of view is favorable. Naturally, when these reflective rings are edge-on to us, they contribute nothing to the planet's brightness and Saturn appears rather faint. Since 1988 the rings have been slowly disappearing from view, and in 1995 they vanish altogether, giving Saturn its faintest magnitude at opposition. The Ringed Planet brightens in each of the following years, partly because its oppositions occur ever closer to Earth and partly because the rings are opening and reflecting more sunlight. By its 2000 opposition Saturn will be 275 percent brighter than it was in 1995!

6 AN INTRODUCTION TO THE STARRY SKY

If you'd never been outside on a clear night, you might think from what you've read so far that there are only seven lights in the sky: the sun, the moon, and the five naked-eye planets. We hope you *have* been outside, though, finding the planets as we've described them. We also hope you've developed some curiosity about their starry backdrop. In this chapter, we'll help you recognize the basic star patterns — enough to get you started on what can be a lifetime's exploration of the marvels of the cosmos.

Odds are you live in a city without dark skies glittering with stars. Our belief is that this isn't really a handicap: The riot of stars that greet the visitor to a dark, rural sky can be tough to sort into the classical constellations. In the city, you can pick out the brightest stars of the brightest constellations. From large parks or dark suburban areas you'll spot some additional stars that will fill out the star patterns we'll show you. From there, you can use a few key constellations to bag more constellations than you probably thought you could ever learn. We're not going to try to show you all the constellations; there are plenty of books that do that. Our aim here is just to get you started. While our charts present the sky you'll see from the city, give yourself a break and scout around for a dark spot in town or on the outskirts. Large parks, especially those near a large body of water, offer enough relief from street lights and brilliant billboards to give you a decent view of the starry sky.

A key to identifying the constellations is to *think big*. Although our charts are reasonable facsimiles of the star patterns, they suggest compact collections of stars. In fact many constellations sprawl across the field of view. The standard way of conveying the extent of constellations is to give their angular size — for example, the Big Dipper stretches twenty-five degrees from end to end. That doesn't mean much to most people. Fortunately, standard issue

for humans is a device that's handy for marking off celestial distances: a hand held at arm's length. Figure 6-1 shows how to use your hand as a gauge of distances. For example, the Big Dipper's bowl is about the size of the average fist held at arm's length; the entire Dipper spans the distance between your thumb and little finger spread wide.

Our star charts include only the stars you can expect to see from a city or suburban location. (Some of the constellation lines suggest the locations of stars that complete a standard figure but are only visible in dark locations.) That doesn't mean you can expect to see all these stars *easily* from the city. Some require a bit of patience to find, but there's also a helpful trick. Because of the way the human eye is constructed, the center of your field of view is not as sensitive to light as the periphery. You will often spot a faint star first by looking one or two finger-widths away from its true location. Using this technique, called *averted vision,* you will be able to detect at least twice as many stars as you thought possible.

Why not start your explorations by looking for that one constellation you've always been especially curious about?

Finding *Your* Constellation

You've probably known for years what your astrological "sign" is. Perhaps you're also aware that your sign, and the other eleven, correspond to the constellations of the zodiac, that band in the sky in which the sun, moon, and planets are always found (Fig. 6-2). This correspondence may actually inspire a few people to go out and look for that special constellation, thereby starting on a new appreciation of the sky — the only benefit of astrology we're prepared to admit!

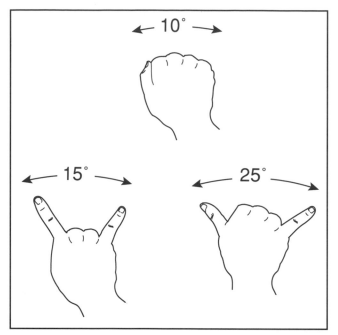

Figure 6-1. *Your hand is a handy yardstick for celestial distances. Your fist at arm's length matches the bowl of the Big Dipper (10°), while the entire Dipper falls within the span of your thumb and little finger spread wide (25°). Stretching your index and little fingers wide forms a measure of the distance across the bottom of the Great Square of Pegasus (15°).*

The zodiac is a kind of celestial bestiary, containing a variety of persons and creatures. Some are mythological, like Gemini, the Twins, others fanciful, like Capricornus — half fish, half goat! The word zodiac stems from the Greek term, *zodiakos kuklos*, "the circle of animals." The names of the individual characters of the zodiac were largely in place in Babylonian times, although the details of the constellations have changed. Table 6-1 lists the zodiacal constellations and the figures they represent. Don't expect to see a recognizable shape in the stars, even from the very best sites: most constellations, particularly those of the zodiac, require a hallucinogenic imagination to discern anything resembling their namesakes.

Although the constellations of the zodiac are a staple of the astrologer, they are not, in fact, uniformly prominent in the night sky; a few are downright invisible to city dwellers. But as you get the hang of finding the planets, you can use them as guides to the region of the sky where your constellation resides. The planets Mars, Jupiter, and Saturn are best suited for this, since Mercury and Venus never stray far from the glow of twilight. Table 6-2 lists the zodiacal locations of the three outer planets for the next eight years.

While we're on the subject of the zodiac, we should point out a fundamental flaw in the system of astrology. What does it mean to say you're born under the sign of Leo? To the founders of astrology, it meant that you were born when the sun was in the constellation Leo, that is,

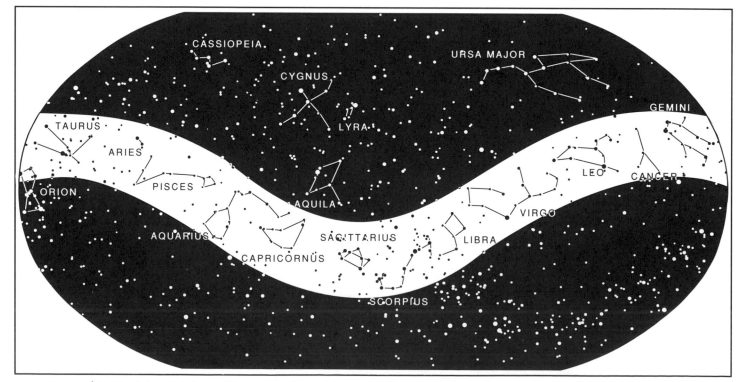

Figure 6-2. *The "signs" of the zodiac are really nothing more than twelve constellations that lie along the ecliptic, the path of the sun, moon, and planets. This map shows the entire celestial sphere, but stars much farther south than the tail of Scorpius cannot be seen from the United States.*

Table 6-1. Constellations of the Zodiac

Aries	Ram
Taurus	Bull
Gemini	Twins
Cancer	Crab
Leo	Lion
Virgo	Virgin
Libra	Scales
Scorpius	Scorpion
Sagittarius	Archer
Capricornus	Sea-Goat

between July 23 and August 23. Nowadays, the sun enters the constellation Leo around August 10 and remains there until mid-September. Yet anyone born between July 23 and August 23 is still instructed to look at Leo's fortune — even if they were born on, say July 25. This discrepancy arises because of a phenomenon known as *precession of the equinoxes*.

As we discussed in the introduction, the earth's spin axis is tilted — that's the reason for seasons, and the reason the sun sets at different points on the western horizon throughout the year. Now the earth also wobbles, or precesses, like a spinning top that's winding down — its axis of rotation traces a circle in the sky once every 25,800 years. As it moves, the place where the earth's axis points neither toward nor away from the sun — an equinox — must occur at different locations in our orbit. For example, about 6,600 years ago (that's one-quarter of 25,800) the equinoxes came at a spot a quarter of the earth's orbit away from where they now occur — meaning that the sun then appeared a quarter of the way around the zodiac. Today the spring equinox occurs with the sun in Pisces, but 6,600 years ago it lay in Gemini. By the way, the location of the spring equinox is known as "the first point of Aries." The name was given to it 2,000 years ago — when the sun *was* in Aries at the spring equinox.

Precession was noticed by the great Greek astronomer Hipparchus in 130 B.C., but astrologers apparently have never become aware of it. Now, as we've said, many of the zodiacal constellations are faint and difficult to discern. So if you can't find your astrological sign, look at the bright side: Thanks to precession, it probably isn't *really* your birth sign, anyway!

With that out of the way, turn to the section that matches the season and start getting acquainted with the rest of the starry sky. Each section is accompanied by a special finder that will guide you to the easiest constellations to spot. More thorough maps covering the sky for the entire year will be found at the end of the chapter.

The Glittering Gems of Winter

Although we typically spend a lot less time outdoors during the cold winter months, the sky at that time of year attracts the attention of many who otherwise don't give the sky a passing thought. Even as we rush from warm car to warm home, we're often compelled to pause and admire the stars that glitter overhead like crystals in a celestial chandelier. If you've wondered why the stars seem brighter in the winter, the answer is simple: They *are* brighter. Table 6-3 lists the brightest stars visible from earth. From the law of averages, you'd expect that one-fourth of them are in winter constellations. Yet nearly half of the brightest stars — seven of fifteen listed — are located in the winter sky. Two of those stars lie in a single constellation, Orion the Hunter. That's the constellation

Table 6-2. Planets in the Zodiac, 1993-2001. Near opposition, the planets Mars, Jupiter, and Saturn guide you to all but one of the zodiacal constellations.

Constellation	Mars	Jupiter	Saturn
Gemini	1993	—	
Cancer	—	—	—
Leo	1995	—	—
Virgo	1997-99	1993	—
Libra	—	1994	—
Scorpius	2001	1995	—
Sagittarius	2001	1996	—
Capricornus	—	1997	1993
Aquarius	—	1998	1994-95
Pisces	—	1999	1995-98
Aries	—	2000	1999
Taurus	—	2001	2000-01

Table 6-3. The Fifteen Brightest Stars

Name	Magnitude	Constellation	
Sirius	-1.46	Canis Major	
Canopus	-0.72	Carina	*
Alpha Centauri	-0.01	Centaurus	*
Arcturus	-0.04	Bootes	
Vega	+0.03	Lyra	
Capella	+0.08	Auriga	
Rigel	+0.12	Orion	
Procyon	+0.38	Canis Minor	
Achernar	+0.50	Eridanus	*
Betelgeuse	+0.50	Orion	
Altair	+0.77	Aquila	
Aldebaran	+0.85	Taurus	
Alpha Crucis	+0.87	Crux	*
Antares	+0.96	Scorpius	
Spica	+0.98	Virgo	

* These inconspicuous or invisible constellations are not labeled on the star charts.

we'll use as our guide to the bright constellations of the winter months. Refer to Fig. 6-3 as you investigate the crystalline winter sky.

Any clear night at the beginning of the year, step outside and face south at about 10:30 P.M. — a swarm of bright stars will greet you. Scan the sky from horizon to zenith (directly overhead). You're bound to notice a distinctive row of three stars. This is the belt of Orion. To the upper left of the belt is a distinctly reddish star — Betelgeuse — and to the lower right is the bright, bluish star Rigel. These stars represent the right shoulder and left knee, respectively, of Orion. Like many star names, these are from the Arabic: *Ibt al Jauzah*, "the armpit of the central one," and *Rijl Jauzah al Yusrah*, "the left leg." Naturally, like any reasonably equipped hunter, Orion has two shoulders and knees; scan to the right of Betelgeuse to find Bellatrix and to the left of Rigel to find Saiph, about where you'd expect the corresponding shoulder and knee. To the right, or west, of Bellatrix is a faint string of stars that just might be visible with averted vision. This is the giant's shield, with which he fends off the rush of Taurus, the celestial bull.

Dangling from Orion's belt is a sword of three faint stars. With averted vision, you may notice that the central star in the vertical trio has an odd misty glow, unlike the pinpoints of light adjacent to it. With a pair of binoculars, you can see that this is not some optical illusion: this is not a star at all, but the Orion Nebula, a vast cloud of dust and gas within which stars are forming. Young, hot stars emerging from this cosmic nursery are what make the

nebula's gases glow. The Orion Nebula lies about fifteen hundred light-years distant — that is, the light we see has taken fifteen hundred years to reach us — and is about seventeen light-years across. Large as that is, it is in fact only the brightest region of a far larger complex that encompasses the entire constellation. One reason the winter stars are so bright is that many are young, and any collection of young stars has a disproportionate number of hot, bright members. Hot, bright stars have short life spans. A star like Rigel may shine for a mere ten million years (whereas our sun has shone for five *billion* years, about half its expected lifetime). In regions of the sky with a more middle-aged population, such stars have long since disappeared. Regions of star birth such as the Orion Nebula are a characteristic of the plane of the Milky Way, the disk of our galaxy, toward which we look in winter. Out of this plane there is little gas and dust from which young stars — and their attendant planets — can form today. Let's turn now to Taurus, a constellation whose stars we might consider "adolescent."

Taurus is a zodiacal constellation whose main stars are easy to find. Orion's belt points up and to the right (northwest) to the reddish star Aldebaran. Aldebaran marks one tip of a V-shaped cluster of stars known as the Hyades. The Hyades mark the face of the Bull; his horns stretch fifteen degrees to the left, or east. Roughly the same distance to the upper right you'll find the Pleiades, the lovely little collection of stars that many people mistake for the Little Dipper. The Pleiades are known as the Seven Sisters, though few people can spot more than six

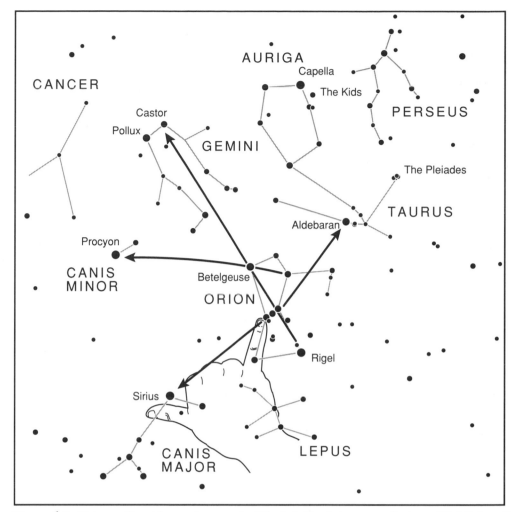

Figure 6-3. *Orion will guide you to several of the bright winter constellations. Highlights of the season include Sirius, the brightest star in the sky, and the compact Pleiades star cluster. The Orion Nebula, a region of star formation in the sword of Orion, deserves a look with binoculars or a telescope.*

stars with the naked eye. A pair of binoculars will reveal perhaps a dozen or so stars, and the cluster totals several thousand.

The Pleiades and Hyades are cousins, both mythologically and scientifically. The Hyades, daughters of Atlas and Aethra, and the Pleiades, daughters of Atlas and Hesperis, are both *open* or *galactic* clusters of stars born of enormous clouds of dust and gas. The Hyades cluster, at a distance of about 140 light-years, is one of the closest to our solar system. The members of this cluster were all born from the same vast cloud of dust, and are known to be traveling together in space. The Pleiades, about three times farther away, is considered a younger star cluster, perhaps 50 million years old. Photographs reveal wisps of dust and gas still surrounding the young stars of the Pleiades, remnants of the primordial cloud from which the cluster formed.

The Pleiades cluster is such a distinctive pattern that its worldwide prominence in folklore should come as no surprise. Among tribes in the Amazon valley its appear-

ance marked the start of the rainy season and the migration of birds. The Guaranis of Paraguay begin their year when the Pleiades make their first appearance in the predawn sky. The Navajo and Blackfeet of North America similarly used the cluster as the basis of a stellar calendar. When the Navajo see *Dilyehe* (Pleiades) low in the early morning (late May), they know it's too late to plant and still be able to harvest before the first frost; the arrival of these stars in the evening sky (late September) serves as a warning that the first frost is near. The principal Navajo deity, Black God, even uses the cluster to adorn his forehead. In Bali and among the Xhosa of South Africa, the Pleiades are consulted when the lunar calendar gets out of synch with the growing season. Many aboriginal peoples of Australia associate a rainy period with the appearance of the Pleiades, and curse the cluster if rain fails to follow it.

The Aztecs of Mesoamerica believed that the Pleiades would herald the end of the world and began their most important ceremony on a date when the cluster — known

Table 6-4. The Fifteen Nearest Stars.

Name	Distance (light-years)	Magnitude	Constellation	
Sun	0.000016	-26.8	—	
Alpha Centauri A	4.4	-0.01	Centaurus	*
Alpha Centauri C	4.3	+10.7	Centaurus	*
Barnard's star	5.9	+9.5	Ophiuchus	
Wolf 359	7.6	+13.5	Leo	
BD +36°2147	8.1	+7.5	Canes Venatici	
Sirius	8.6	-1.46	Canis Major	
Luyten 726-8	8.9	+12.5	Cetus	
Ross 154	9.4	+10.6	Sagittarius	
Ross 248	10.3	+12.2	Andromeda	
Epsilon Eridani	10.7	+3.7	Eridanus	
Luyten 789-6	10.8	+12.2	Aquarius	
Ross 128	10.8	+11.1	Virgo	
61 Cygni	11.2	+5.2	Cygnus	
Epsilon Indi	11.2	+4.7	Indus	
Procyon	11.4	+0.38	Canis Minor	

* These two stars — plus a third component not listed — orbit one another as part of a single star system.

to them as *Tianquiztli*, "market place" — crossed overhead at midnight. At the end of their fifty-two-year calendrical cycle, when the dates of their secular and sacred calendars coincided, priests watched anxiously as the cluster approached the zenith. If the Pleiades stopped moving as they passed overhead then the world would come to an end; otherwise it would continue on through another fifty-two-year cycle.

Before we turn from Taurus, find the star that marks the tip of the bull's lower horn, above Orion. About one degree (the width of a finger tip) to the upper right lies the remnants of a star that, over nine hundred years ago, shone brightly enough to be seen in the daytime. Now invisible to the naked eye, this debris is the Crab Nebula, an expanding cloud of gas that marks the site of the spectacular supernova of A.D. 1054. We'll return to that famous event in Chapter 8.

Our next stop in the winter sky is a pentagon of stars lying above the horns of Taurus — Auriga, the Charioteer. The most obvious member of this constellation is the bright star Capella. To the ancient Greeks, Capella represented Amalthea, a she-goat that suckled the exiled Zeus (see Chapter 5). We mention this because it's about the only way to make sense of the appellation given to the trio of stars just beneath Capella — The Kids! For the beginning observer, Auriga isn't a terribly interesting constellation. A small telescope reveals a number of star clusters like the Pleiades and Hyades, though much more distant and hence smaller and fainter.

Returning now to Orion, we'll trace the line of his belt away from Taurus to the brightest nighttime star visible

from earth. This is Sirius, whose brightness stems largely from its proximity to earth, a relatively nearby 8.6 light-years. Sirius (the name probably stems from the Greek for hot or scorching) is the Dog Star, the brightest star in Canis Major, the Great Dog. This star's appearance in the east just before sunrise in early August signals the onset of the hottest days of the year — the Dog Days. To the ancient Egyptians, this first appearance of Sirius (called *Sothis*) was of great importance for it signaled the start of the vital Nile flood, a coincidence that led them to the discovery of a 365-day year by 2,700 B.C. Canis Major is one of the few constellations whose rough outline does suggest the shape of its namesake. Most of these stars will be visible from suburban skies, though the dog's head is made up of fainter stars. The full constellation is about fifteen degrees long.

Animals seem to come in pairs in the heavens, and the Great Dog has a lesser companion to the north. Trace a line eastward (left) through the shoulders of Orion until you come to the bright star Procyon. This star is Canis Minor's main claim to fame, though beginning stargazers may also appreciate its simplicity: only one other star is required to complete its standard figure! Canis Minor, along with the Greater Dog, is often considered one of the hounds Orion took on his hunts. Procyon makes it into the list of the top ten brightest stars, and also ranks fifteenth in a list of the nearest stars (see Table 6-4).

Our final winter stop is another constellation of the zodiac — Gemini, the Twins. The pair of bright stars that marks the heads of the Twins can be found along a line extended from Rigel through Betelgeuse. The brighter of

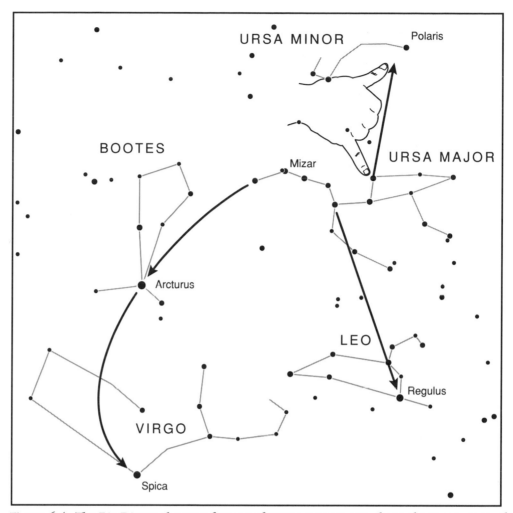

Figure 6-4. *The Big Dipper, the most famous of spring star groups, shows the way to several constellations. Mizar and Alcor, in the Dipper's handle, also serve as a test of your vision!*

the two stars is Pollux; his twin Castor lies nearer the Pole Star. The constellation is about equal to Orion in length, extending toward the Hunter from the twin stars. Castor and Pollux would stand out in any other season; they pale before the surrounding winter stars, however. This constellation is of interest mainly to those naked-eye astronomers "born under" the sign of the twins. Of historical interest: Gemini is the constellation in which Clyde Tombaugh of Lowell Observatory discovered Pluto in 1930.

Before we leave the winter sky, it's interesting to observe a characteristic of stars that is often unnoticed, but fairly obvious when pointed out: their colors. The winter sky contains some fine contrasts in this regard. Compare the stars Betelgeuse and Rigel in Orion, for example. Betelgeuse is clearly reddish, while you may detect a hint of blue in Rigel. Aldebaran in Taurus is another colorful star, running perhaps a bit more toward the orange than Betelgeuse. The colors of stars are an important clue to their physical nature since it reveals their surface temperatures. The bluest stars are the hottest; Rigel's surface temperature runs about 12,000° C. The cooler the star is,

the redder it appears; Aldebaran is "only" about 4,000° C. The white and blue-white stars are stars in the prime of life, but redder stars like Betelgeuse and Aldebaran are nearing the end of their lives. They have reached the so-called "red giant" phase, a period marked by a dramatic expansion that spreads their energy over a larger surface area and reduces their surface temperatures. More massive than our own sun, Betelgeuse actually qualifies as a red supergiant. If placed at the sun's position in the solar system, Betelgeuse would engulf all the planets out to Jupiter!

The Stars of Spring

After the bright stars of winter, the spring sky may seem a bit drab. But there are compensations. While the spring stars are generally not as bright as those of winter, the temperatures tend to be more tolerable for those who want to linger outdoors. The major constellations are quite easy to identify, since we have a great guide constellation. Figure 6-4 gives an overview to finding your way around the spring sky.

The key to the spring stars is a constellation that isn't a true constellation. The familiar Big Dipper, also known as the Plow in Europe, is really just part of Ursa Major, the Great Bear of the north. Such a prominent grouping of stars that forms only part of a true constellation is known as an *asterism*. The Big Dipper lies relatively close to the north celestial pole, and is one of a handful of *circumpolar* star groups that never set. For the purpose of finding the spring constellations, the best time to look for the Big Dipper is about 10:30 P.M. local daylight time at the beginning of April, two hours later in May. Face south to locate the Big Dipper, whose stars are visible from all but the most brightly lit urban sites. Hold your hand nearly overhead with your thumb and little finger stretched as far apart as possible; this is the extent of the Dipper. It should be fairly easy to identify the distinctive bowl and curving handle. (Observers in the southern United States — especially Hawaii! — may find it more easily facing north, since it appears lower in the sky; bear in mind, though, that it will appear upside down compared to Fig. 6-4.)

Once you've spotted the Big Dipper, take a good look at the middle star of the handle. This star, Mizar, has a faint companion that can be picked out by those with keen eyesight. The companion, Alcor, is another candidate for the averted vision trick. If you can't see Alcor, don't rush to your optometrist — try a pair of binoculars. The pair is sometimes called "The Horse and Rider," and medieval Arab writers referred to it as "The Test." This is a *double star*, two stars that appear as one from our viewpoint on earth.

Since we told you all about galactic star clusters in the winter section, it's worth mentioning that most of the stars of the Big Dipper are members of the closest known cluster, lying about seventy-five light-years off. Interestingly, stars in a large volume of space that includes the sun appear to be moving in the same general direction as the stars of the Big Dipper cluster. Perhaps, five billion years ago or so, our sun was born in a region of the same vast cloud of gas that gave rise to the stars of the Big Dipper.

Turn north now to find a single star. The two stars of the Dipper's bowl farthest from the handle are known as the Pointer Stars; they form a line which, if extended about a hand-span northward, passes near a single moderately bright star, Polaris. Polaris, the North Star or Pole Star, lies almost directly above the earth's north pole and is the pivot around which the celestial vault revolves. The Pawnee of North America called it "the star that does not walk around," and groups in Europe and Asia referred to it as the "nail of the world" or "pillar of heaven." Polaris marks the end of the handle of the Little Dipper, Ursa Minor (the Lesser Bear). Most of this constellation's stars are too faint to see from city sites, though the stars corresponding to the Big Dipper's Pointers are visible to most observers.

While the North Star seems unmoving in the night sky, it isn't precisely on the north celestial pole. A long-exposure photo will show a noticeable trail, evidence of Polaris's motion. The Pole Star is a symbol of steadfastness and reliability, but as we noted earlier, the earth wobbles slowly and this precession causes the earth's axis to move across the sky. When the Pyramids were being built, the star Thuban in Draco was the pole star; 12,000 years ago, the bright summer star Vega was near (though not at) the north celestial pole. Currently, the earth's axis is moving still closer to Polaris; its closest approach comes in the year 2102.

Let's turn back south to find some constellations that you have a better chance of seeing in their entirety.

Again, we'll start at the Big Dipper, whose arcing handle also serves as a pointer. Extend an imaginary curve away from the bowl to the southeast until you spot a bright, yellowish star. This is Arcturus, the brightest star in the spring sky. (Don't be led astray by the bright *planet* Jupiter, which is in the vicinity until the end of 1994.) Arcturus is the brightest star in Bootes, the Herdsman. Bootes's rather faint stars form an elongated kite, about a hand-span high, extending north from Arcturus.

From Arcturus speed on to Spica, the next bright star along the same arc that brought you to Arcturus. Spica is the brightest star in Virgo, one of the zodiac's more conspicuous constellations. Virgo was associated with the harvest, and the name Spica refers to an ear of wheat. Virgo is one of the largest constellations, nearly two hand-spans across. It happens that the spring sky gives us a view out of the plane of our home galaxy, the Milky Way (see Fig. 6-6). The dense clouds of gas and dust lying within the Milky Way block our view of the extra-galactic universe. Galaxies are scattered throughout the spring sky. But when we come to Virgo, we find an astounding concentration of galaxies known as the Virgo Galaxy Cluster. Over three thousand member galaxies have been found with professional telescopes. Amateur telescopes can reveal perhaps a hundred of these in the region about a hand-span north of Spica. Although invisible to the naked eye, it's remarkable that they can be seen at all since they lie fifty *million* light-years away. None of the galaxies in the cluster itself is terribly photogenic. Still, it's worth noting that this seemingly empty patch of sky teems with literally trillions of distant suns.

Our final spring target is Leo. If you haven't already made out its distinctive shape, return to the Big Dipper. The side of the bowl near the handle acts as yet another set of pointers, this time south to a bright star halfway between the Big Dipper and the southern horizon. This is Regulus, the "little king" of Leo, the Lion. Leo's most distinctive feature is "the sickle," a backwards question mark that rises from Regulus. A right triangle of stars represents the lion's haunches, about a hand-span to the left, or east.

To the right of the sickle is Cancer, one of the least conspicuous constellations of the sky. The only real interest this constellation may hold for the beginning astronomer is the star cluster at its heart. This object, with the

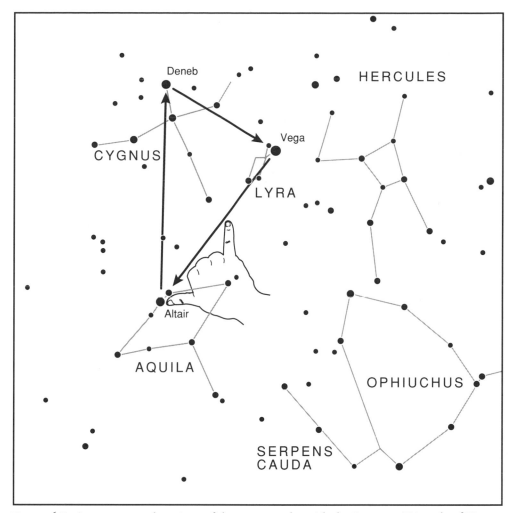

Figure 6-5. *Start your explorations of the summer sky with the Summer Triangle of Vega, Altair, and Deneb. Farther south lie two distinctive constellations, Scorpius and Sagittarius.*

whimsical name of "The Beehive," is a naked-eye object out in the countryside. It may be enjoyed with binoculars from a suburban site, but locating it will be a challenge.

The Summer Sky

Summer is a bit frustrating for stargazers. The temperatures are congenial, but the nights start late and tiny buzzing vampires can be a nuisance in many regions. The hours don't need to be a problem, at least while you're on summer vacation. While on the road, plan to do some nighttime driving; you can then pull off the interstate for a brief break on a dark country road, where the splendor of the summer sky is at its very best. The summer sky lacks a single constellation that can serve as a guide to the others. But, as Fig. 6-5 shows, there are features that will help you find your way around. And, after all, on those balmy summer nights there's no great rush to learn the sky — just slather on the insect repellent, relax on your lawn chair, and enjoy it!

Rather than having you search for a particular constellation, we invite you to simply face east and scan the sky for a bit. You'll soon spot a bright star nearly overhead. A little lower is another star not nearly as bright. Scan to the south (right) and you'll find a third bright star. Together, these three stars, each one the brightest of its own constellation, make up the Summer Triangle.

The brightest of these three stars is Vega, in Lyra the Lyre. The other four stars of this lovely constellation are easy to pick out from a dark site since they are fairly bright and lie nearby. This is the lyre that the luckless Orpheus used to charm the guardians of Hades so that he could rescue his beautiful wife Eurydice from the realm of the dead. (He lost her when, contrary to instructions, he looked back too soon to see if she was still behind him.) The Chinese tell a lovely tale in which Vega is Chih-nu, a divine princess who fell in love with a mortal, a poor herdsman no less. Her father, the sun god, didn't think this a proper match and placed them in the sky, separated by a heavenly river. Once a year, however, a sympathetic

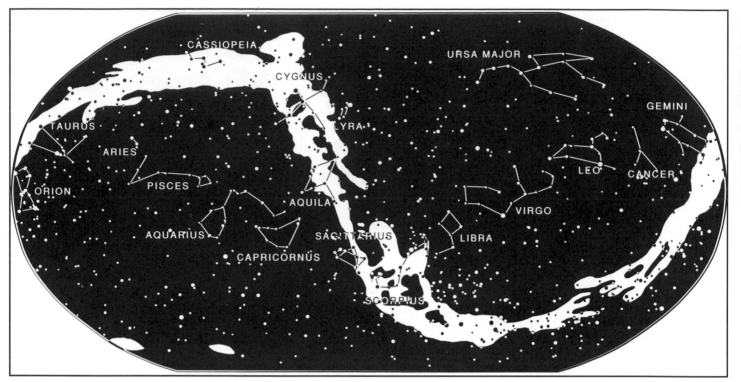

Figure 6-6. *The faint band of the Milky Way (white) represents the dense population of stars in the plane of our galaxy. Intervening clouds of dust give it its ragged appearance. Follow the Milky Way from Cygnus and Aquila to the southern zodiac constellations Sagittarius and Scorpius.*

flock of magpies form a bridge between the lovers, who are thus briefly reunited.

Ch'ien-niu, the herdsman of that tale, was known to westerners as Altair, the second brightest star of our stellar triangle and the farthest from Vega. In the western bestiary, Altair lies at the head of Aquila, the Eagle. This is another one of those rare constellations that bears at least a vague resemblance to its namesake. The eagle flies northeast with broad wings over a hand-span across.

Deneb, the third star of the Summer Triangle, is also part of an avian constellation. Deneb marks the tail of Cygnus, the celestial swan, which is flying in the opposite direction of Aquila. Each of the swan's wings are nearly a hand-span in extent, although in the city you may only be able to trace them about half that distance. The neck is equally long, with the head marked by a somewhat faint star not far from the southern end of Lyra.

If you have a geometric turn of mind, it may have occurred to you that if in the winter we looked into the plane of the Milky Way, then out of the plane in spring, summer would give us another view of our galaxy. That is indeed the case, and in fact summer brings the very best views. Cygnus the Swan is flying right down the Milky Way, the glowing river of stars that represents the stream that separates the mythical Chinese lovers. If you trace the Swan's imaginary flyway southward, you'll come to the very heart of our galaxy.

As you face south, you'll see a reddish star, Antares, the brightest star in Scorpius, the Scorpion. This constella-

tion is reasonably bright and distinctive, and should be easy to trace in its entirety. Antares means "rival of Mars," a name whose wisdom is particularly apparent when the Red Planet passes by. The best opportunity for comparison comes in 2001, when Mars is near opposition in Scorpius. (See Figure 4-7 in Chapter 4.)

Trace the Scorpion's figure to the end of its tail, then scan to the left — you'll come upon a group of stars that forms a teapot about the size of your open hand. This teapot is part of a group of stars that the classical ancients thought of as Sagittarius, the Archer. The spout of the teapot can be thought of as pouring almost directly into the heart of the Milky Way. Study this constellation well. If you make a "star stop" in the countryside this summer, you may be surprised by how difficult it is to see this distinctive shape when it's embedded in the glow of the Milky Way. Since we're looking back into the plane of the Milky Way, thick with clouds of gas and dust, we also have a view toward many regions of star formation. Sagittarius is particularly rich in this regard. A pair of 7 x 50 binoculars can be used to scan the Sagittarius region for dense clouds of stars and nebula, which appear as faintly milky patches of light — if you're away from bright city lights.

The summer Milky Way is worth a special trip out to the countryside for a good look (Fig. 6-6). Again, a pair of binoculars is a great help, but not absolutely necessary for a rewarding experience. To truly appreciate the sight, you have to give your eyes time to adapt to the darkness. Two changes take place. First your pupils dilate to the maximum size, allowing more light into your eyes; this takes only a few minutes. There's also a chemical change that makes the receptors more sensitive to light; this "dark adaptation" takes about thirty minutes to complete. It's important to stay away from lights (except red lights) since just a brief glimpse at a dome light or car headlight will immediately reverse the process, and you'll have to start all over again. As your eyes slowly become dark adapted, you'll be surprised at the variations of light and dark that become apparent in the Milky Way. In particular, note the region near the head of Cygnus; this is the beginning of the Great Rift, an elongated dark patch that splits the Milky Way down to the horizon. Interestingly, villagers in modern Peru count among their constellations animal shapes formed by dark patches in the southern Milky Way — probably an astronomical relic of the great Inca empire.

As we said earlier, the Milky Way is a sort of celestial flyway for Aquila the Eagle and Cygnus the Swan. It runs from above Cygnus down to Sagittarius. The name Milky Way comes from the Latin *via lactea*; the Greeks called it *galaxias kuklos,* the milky circle, which is where we get the generic term galaxy. Many cultures regarded the Milky Way as the river of heaven — we saw this reflected in the Chinese myth. Some early natural philosophers speculated that the glow of the Milky Way was the product of innumerable unresolved stars, but it wasn't until Galileo examined the region with his telescope that the matter was settled. It was over a hundred years, however, before anyone suggested that the glowing circle was actually our home in the cosmos. In the late eighteenth century, William Herschel attempted to map the galaxy by counting stars and correctly guessed the structure of the Milky Way. At the beginning of the twentieth century, Harlow Shapley charted old star clusters and recognized that they were distributed in a roughly spherical pattern. He also realized that the center of the sphere was the center of the galaxy.

Today, astronomers reckon that the main disk of the galaxy is about a hundred thousand light years across, with our solar system lying about thirty thousand light years from the center — in the suburbs, as it were. Even at that great distance, the glow of the concentration of stars at the galactic center would be as bright as the full moon, if our view were not blocked by intervening dust and gas. That view is blocked only to human eyes; radio telescopes have been used to study the radio waves that pierce the dusty veil. Such studies indicate that stupendous amounts of energy are being released at the galactic center. One theory is that enormous stars are being born and exploding in a ferocious "starburst" that produces as much energy as 30 million suns. A somewhat more intriguing idea is that a black hole several million times more massive than the sun lies at the galactic center, gobbling up matter. Material spiraling into such a black hole would give off vast amounts of radiation before disappearing for eternity.

So, as you reflect upon the silent beauty of the Milky Way, imagine the titanic roar of radio energy emitted at the invisible heart of our galaxy!

The "Watery" Autumn Sky

As the earth revolves around the sun, the planet soon turns its night side away from the galactic center toward a region of the sky that seems almost starless by comparison. As in spring, we're largely looking away from the galaxy's disk. The shortening days make the Summer Triangle a prominent feature of the evening sky (leading some wags to call it the "fall triangle"). Its prominence is no doubt enhanced by the dearth of prominent constellations in the fall sky. In particular, the "watery" constellations of the zodiac — Capricornus the Sea-Goat, Aquarius the Water Bearer, and Pisces the Fishes — are as faint as they are famous. Nevertheless, the fall sky does have some treats, including one of the year's very best. Figure 6-7 illustrates the highlights of fall and the best routes for finding them.

Our first stop is a regal constellation, Cassiopeia. To find this celestial queen, face north about 10:30 P.M. in mid-September. If you have a clear view to the horizon, you may see the Big Dipper skipping along, low in the sky. Use the Pointers to find Polaris, but keep going until you find a group of stars that form a rather lopsided figure 3 nearly overhead. (Depending on the time of year and hour, Cassiopeia variously resembles an M, a W, an E, or a 3.) As usual, this constellation looks small on our maps, but is rather large in the sky; it covers roughly the space between your index and little fingers held at arm's length. Cassiopeia was the matriarch of a legendary family that included Cepheus and Andromeda, two more autumn constellations. Andromeda married Perseus, rider of Pegasus, the winged horse — both also represented in this season's stars. Perseus won Andromeda's hand by rescuing her from Cetus, the sea-monster (or whale), one of the faint water constellations of fall. Cassiopeia, it should be noted, lies embedded in the Milky Way, along Aquila's flyway.

In addition to her mythical notoriety, Cassiopeia serves the useful function of guiding us to the Great Square of Pegasus. Face south now and lean back to spot Cassiopeia, now a sloppy E. Look at the second-lowest leg of the E. This line, extended southward, is a pointer to the Great Square, which rides high in the sky. The Square is about the width of Cassiopeia, but its stars are bright and it should be an easy target. A square doesn't much suggest a horse, of course. To see Pegasus turn northeast, note that the left, or west, edge of the square holds the equestrian features. Extending from the upper left (southwest) corner is the horse's neck and face; the forelegs are kicking from the lower left (northwest) corner. If you want to find the wings of Pegasus, however, you're on your own!

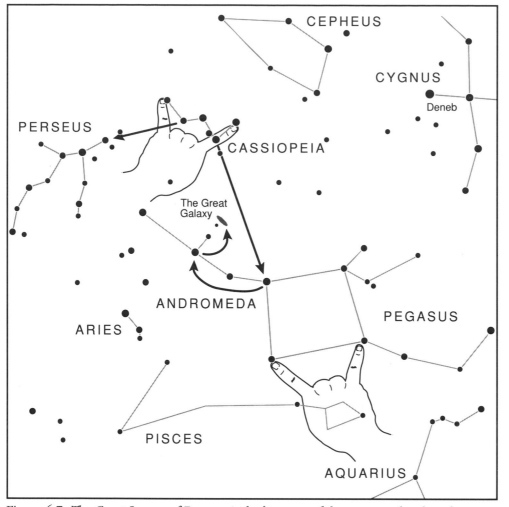

Figure 6-7. *The Great Square of Pegasus is the keystone of the autumn sky, though you may need Cassiopeia's assistance to find it. Hop along Andromeda's chains to locate the jewel of the fall sky: the Great Galaxy.*

We turn now to a constellation that holds one of the sky's most splendid naked-eye sights. Locate the northeast corner of the Great Square. This star is properly considered a member of the constellation Andromeda, the Chained Maiden left out as fish food for monstrous Cetus. In a dark sky you might discern a V of stars extending diagonally from the square; in the city, you'll probably see only the southerly leg of this V. Look two stars over from the square's corner. Now look for a star just above that second star. Continue in a line an equal distance, looking for a third, fainter star. If you can locate this third star, your skies may be dark enough for you to discern a milky patch of light; if not, try binoculars. That smudge represents light that started its journey to your eyes before humans walked the earth. The milky patch is the Great Galaxy in Andromeda, over two million light-years (some 13 thousand quadrillion miles!) from our own galaxy. From dark country skies, the galaxy is quite noticeable as a small glowing cloud of light, making it easily the most distant object that can be seen with the naked eye. Like the Milky Way, this

sight is worth a trip to the countryside. (By the way, the sky we're describing is on display in the early morning hours of mid-August, when the Perseid meteor shower offers yet another good reason to drive out of town; see Chapter 7 for more.) The Andromeda Galaxy lets us see from the outside what we saw from the inside in summer: a system of hundreds of billions of stars forming an "island universe" in the vastness of intergalactic space.

The Andromeda Galaxy was, of course, known to ancient stargazers who had little notion of its true nature. The first record of it is from A.D. 905. For centuries it was known as the Andromeda Nebula, and placed in roughly the same class as the Orion Nebula. To Arabian astronomers, it was the Little Cloud. (The word nebula, by the way, is Latin for cloud.) With telescopes, astronomers eventually saw that the Andromeda Nebula was a member of a class of "spiral nebulas," and one fairly popular view in the nineteenth century held that the swirling spiral nebulas were solar systems at the earliest stage of formation. In 1912 Henrietta Leavitt of Harvard College Observatory made an important discovery about a class of stars called Cepheid variables. These stars changed brightness in cycles that revealed their true luminosity. The slower they went through their variations, the brighter they were. By comparing a star's intrinsic luminosity to its apparent brightness, astronomers can of course determine the star's true distance. In 1924 Edwin Hubble discovered Cepheid variables in the Andromeda Nebula and used them to prove that it is a galaxy in its own right, far beyond and outside our own. Hubble thus took the Copernican revolution a step farther, showing that our galaxy was just one of many spiral galaxies sprinkled through the cosmos.

We've briefly scanned just a handful of constellations. With just a little effort, you should be able to spot these star groups. In time, they'll be familiar companions whose arrivals you'll anticipate with the change of seasons. On the following pages are star charts that will help you fill in the gaps in the sky, giving you more to enjoy as you watch the planets wander through the fixed stars that fill the firmament.

Using the Star Charts

The six star charts on the following pages will round out your introduction to the starry sky. We've included just the stars you can expect to see from a reasonably dark urban site. However, we've also included all the lines used in standard representations of the constellations. This means that some of these lines seem to lead nowhere. Perhaps with averted vision in the city — and certainly in the country — you'll see that stars do indeed lie at the ends of such lines.

To use the charts, simply hold them overhead so the direction on the map matches the direction you're facing. Of course, you'll need a flashlight to see the charts outside, but remember that bright lights decrease your eyes' sensitivity to faint starlight. To save your hard-earned night vision, cover the lens of the flashlight with several layers of red paper, or just a couple of layers of brown shopping-bag paper.

One final point: While we encourage you to stargaze in dark, country skies, our charts will probably do you little good if you're under a dark, star-studded sky. Do your initial stargazing from your backyard or a suburban park, where the glow of lights washes out the faintest stars.

January and February Stars

Early January 10:30 P.M.
Mid January   9:30 P.M.
Early February 8:30 P.M.
Mid February   7:30 P.M.

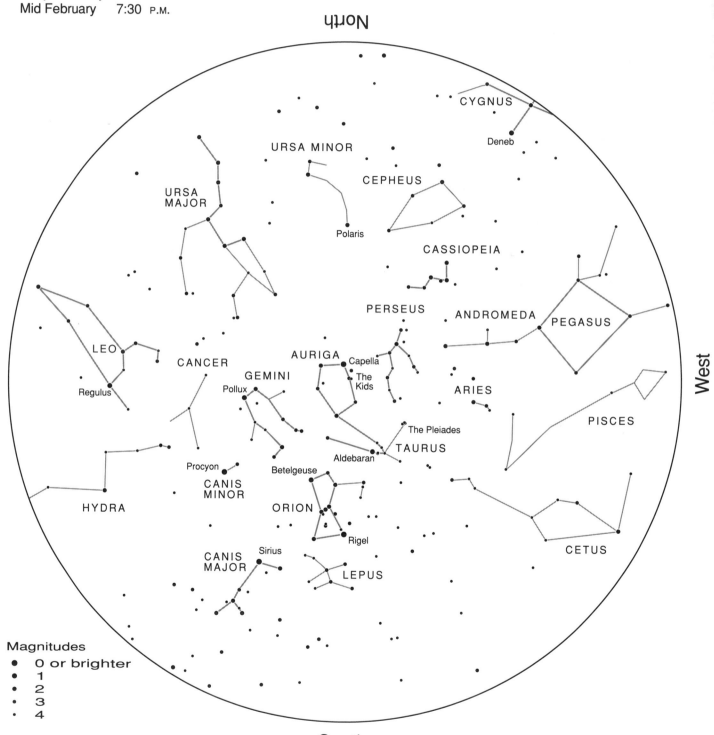

North

East

West

South

CYGNUS
Deneb
URSA MINOR
CEPHEUS
URSA MAJOR
Polaris
CASSIOPEIA
PERSEUS
ANDROMEDA
PEGASUS
LEO
CANCER
AURIGA
Capella
The Kids
ARIES
GEMINI
Pollux
Regulus
PISCES
The Pleiades
TAURUS
Aldebaran
Procyon
Betelgeuse
CANIS MINOR
ORION
HYDRA
Rigel
CETUS
Sirius
CANIS MAJOR
LEPUS

Magnitudes
● 0 or brighter
● 1
● 2
· 3
· 4

March and April Stars

Early March	11:30	P.M.
Mid March	10:30	P.M.
Early April	10:30	P.M.
Mid April	9:30	P.M.

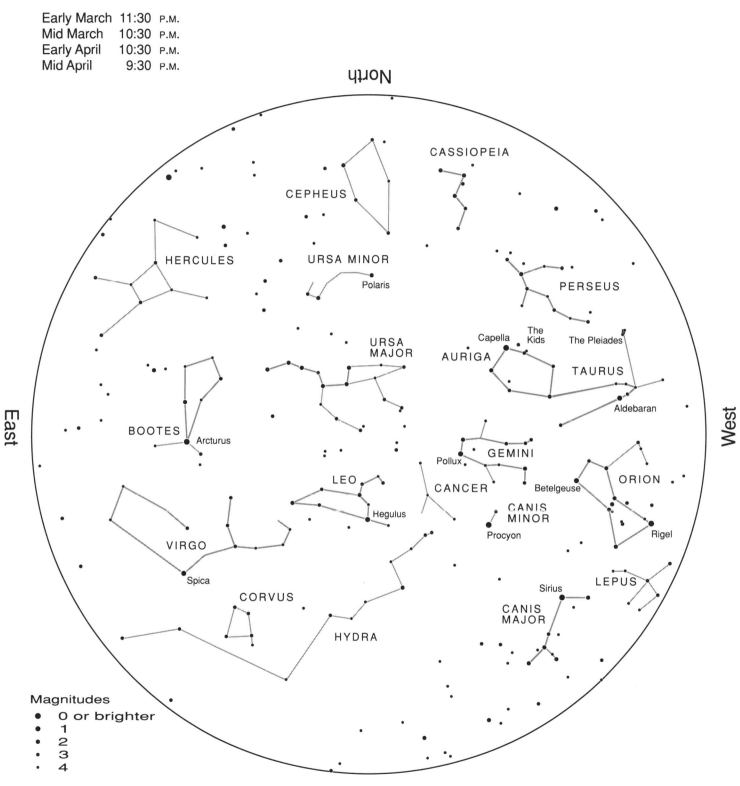

Magnitudes
- 0 or brighter
- 1
- 2
- 3
- 4

May and June Stars

Early May 12:30 A.M.
Mid May 11:30 P.M.
Early June 10:30 P.M.
Mid June 9:30 P.M.

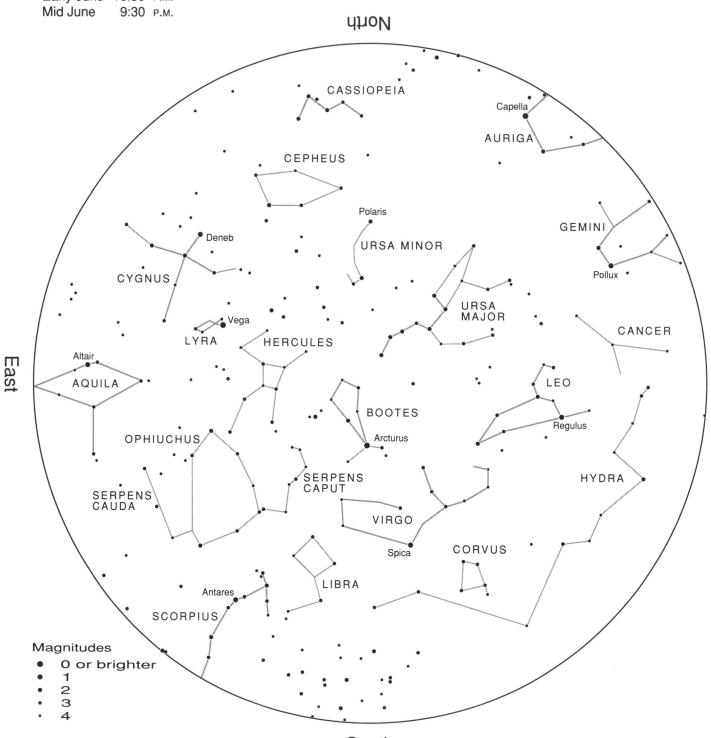

July and August Stars

Early July	12:30	A.M.
Mid July	11:30	P.M.
Early August	10:30	P.M.
Mid August	9:30	P.M.

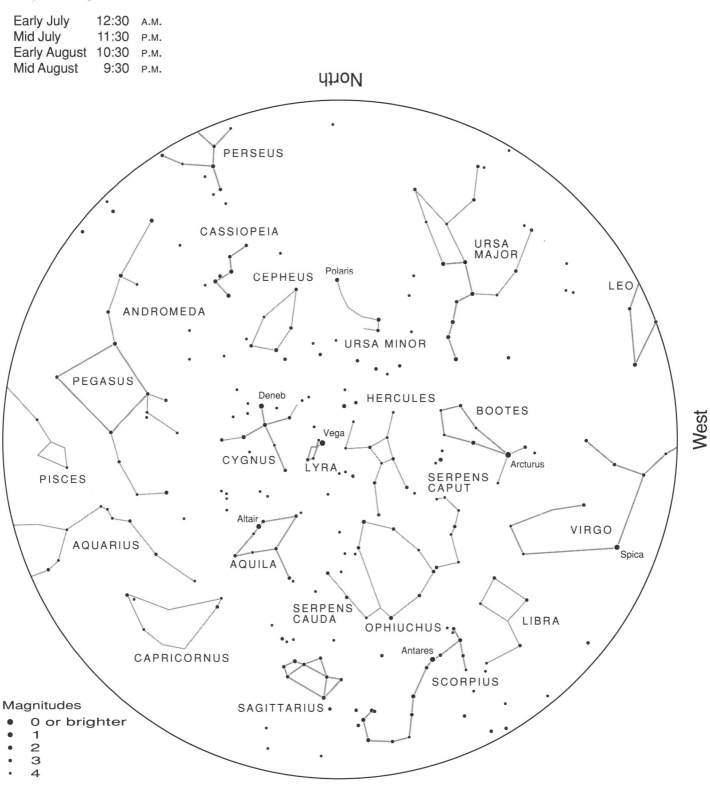

North

East

West

South

PERSEUS

CASSIOPEIA

CEPHEUS

Polaris

ANDROMEDA

URSA MAJOR

LEO

URSA MINOR

PEGASUS

Deneb

HERCULES

BOOTES

Vega

CYGNUS

LYRA

SERPENS CAPUT

Arcturus

PISCES

Altair

VIRGO

Spica

AQUARIUS

AQUILA

SERPENS CAUDA

OPHIUCHUS

LIBRA

Antares

CAPRICORNUS

SCORPIUS

SAGITTARIUS

Magnitudes
- ● 0 or brighter
- ● 1
- ● 2
- ● 3
- · 4

September and October Stars

Early September   11:30 P.M.
Mid September   10:30 P.M.
Early October   9:30 P.M.
Mid October   8:30 P.M.

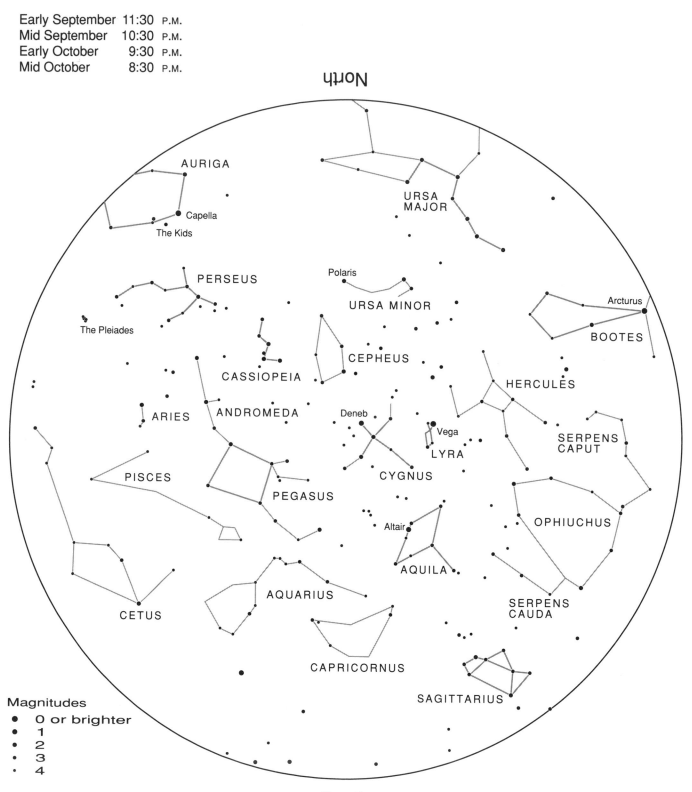

North

East

West

South

AURIGA
Capella
The Kids
URSA MAJOR
PERSEUS
Polaris
The Pleiades
URSA MINOR
Arcturus
BOOTES
CEPHEUS
CASSIOPEIA
HERCULES
ANDROMEDA
ARIES
Deneb
SERPENS CAPUT
Vega
PISCES
LYRA
CYGNUS
PEGASUS
OPHIUCHUS
Altair
CETUS
AQUILA
AQUARIUS
SERPENS CAUDA
CAPRICORNUS
SAGITTARIUS

Magnitudes
● 0 or brighter
● 1
● 2
● 3
· 4

Color Plate 6. *The Voyagers also showed us the splendor of Saturn. Voyager 2 took this image from a distance of 21 million miles in 1981. Like Jupiter, the planet's rapid rotation and fluid interior give it a distinctly flattened appearance. Three of Saturn's satellites (Tethys, Dione, and Rhea) are visible below the planet; the dark spot in the southern hemisphere is the shadow of Tethys. Photo by NASA/JPL.*

Color Plate 7. *The Pleiades, a cluster of hot young stars about 410 light-years away, is one of the treats of the winter sky. Born some 50 million years ago from the same interstellar cloud, some cluster stars still retain a veil of dust that reflects their light. Photo by Hale Observatory.*

Color Plate 8. *Long exposures with a stationary camera reveal the apparent motion of the sky through "star trails." The bright star nearest the center of the whirl is Polaris, the North Star. The dome of the Canada-French-Hawaii telescope on Hawaii's Mauna Kea is silhouetted. Photo by Fred Espenak.*

Color Plate 9. *The next time you find yourself in dark skies look for the Milky Way, the faint band of light that marks the plane of our home galaxy. This picture shows the summer Milky Way from Cygnus to Sagittarius. The bright pale orange "star" near the horizon is Mars, located in the handle of the Sagittarius teapot. Look for the constellations of the Summer Triangle; compare this photo to the all-sky map in Chapter 6 (Figure 6-6). Photo by Fred Espenak.*

Color Plate 10. *The Great Galaxy in the constellation of Andromeda is easily the most distant object visible to the naked eye. A spiral galaxy similar to our own, its light takes 2.2 million years to reach us. Photo by NOAO.*

November and December Stars

Early November 10:30 P.M.
Mid November   9:30 P.M.
Early December 8:30 P.M.
Mid December   7:30 P.M.

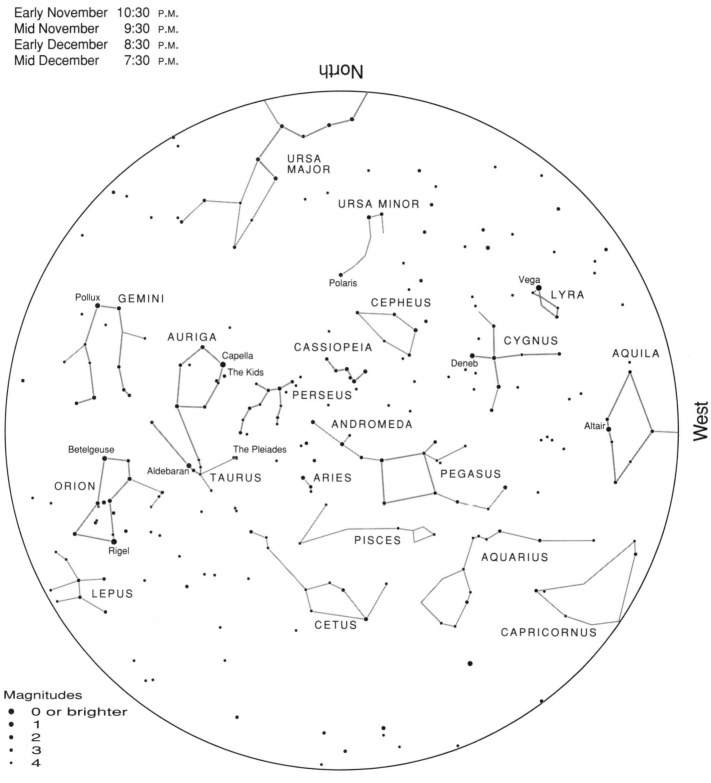

North

East

West

South

Magnitudes
- 0 or brighter
- 1
- 2
- 3
- 4

7 METEORS AND METEOR SHOWERS

As you spend more time under the night sky, sooner or later you'll notice a "shooting star," a streak of light that flashes across the sky in less than a second. This is a meteor, a piece of celestial debris burning up in earth's atmosphere after being swept up by the earth in its travels around the sun. Many meteors are quick flashes detected out of the corner of your eye, but some last long enough for you to track their brief course across the sky. Now and then a meteor will truly light up the night, blazing brighter than Venus and leaving in its wake a dimly glowing train that may persist for minutes. Under a dark sky, you can expect to see between two and seven meteors each hour any night of the year.

But several times during the year the earth encounters swarms of small particles, and the number of these celestial fireflies increases noticeably. The result is a *meteor shower*, during which observers may see dozens of meteors every hour. Concentrations of material within the swarms may produce better-than-average displays with rates of a hundred or more every hour. And rarely we're treated to truly spectacular displays that produce hundreds or thousands of visible meteors for a brief period (Fig. 7-1). The science of meteor astronomy began with just such a meteor "storm," and chances are this same storm will make a return appearance in the late 1990s. So we'll start with a look at the usually unimpressive meteor shower that spawns it — the Leonids of November.

The Story of the Leonids

As a look at Table 7-1 reveals, the Leonids rank among the year's least impressive showers, producing a dozen or so meteors each hour for a couple of nights in mid-November. Backtracking along each meteor's visible trail shows that they all seem to radiate from a point in the constellation Leo, within the "backwards question mark" that rises from the bright star Regulus. This *radiant* is an effect of perspective. The particles that become meteors travel in roughly parallel paths as the earth plows into them and, just as parallel railroad tracks seem to converge on a distant vanishing point, so too do the meteors (Fig. 7-2). All meteor showers are named for the constellation in which their radiant is located.

The tale of the Leonids begins late on November 11, 1799, at the Venezuelan encampment of explorer Alexander von Humboldt. He noticed the meteors around midnight, but as time passed they fell with greater frequency. By 2:30 A.M., he wrote, there was "no part of the sky so large as twice the moon's diameter not filled each instant by meteors." He wasn't the only one to witness the event. A Florida observer wrote that the meteors were "at any one instant as numerous as stars," and in England the shower was described as "sublimely awful." The Leonids, at this time not even recognized as a meteor shower, then lapsed back to their usual trickle until 1831, when a few European observers counted over a hundred shooting stars each hour on the morning of November 13.

But the best was yet to come. In 1832 meteors rained through the predawn sky from the Ural Mountains to the eastern shore of Brazil — an incredible display amounting to twenty thousand shooting stars per hour! It's difficult to imagine a meteor shower getting any better than that, but in 1833 the Leonids outdid themselves. Along the east coast of North America, from Canada to Mexico, anyone under a clear sky in the hours before dawn saw hundreds of meteors every *minute*, a rate of perhaps fifty thousand per hour. An Annapolis observer described the meteors as falling "like snowflakes." A planter in South Carolina wrote: "The scene was truly awful, for never did rain fall much thicker than the meteors fell towards the earth; east, west, north, and south, it was the same." Some accounts of the storm describe a black cloud overhead from which the

Figure 7-1. *The great meteor storm of 1833 lights up the sky above Niagara Falls, New York. The cluster of debris that caused it created similar showers in 1799, 1866, and 1966, and will return to earth's vicinity in the 1990s. Courtesy of Griffith Observatory.*

meteors shot — an illusion caused by the hard-to-see meteors closest to the radiant. Others noticed that the meteors seemed to stream from the constellation Leo and that this radiant moved with the stars.

In the days that followed, wild theories involving electrified air or flammable gases filled the popular press. Accounts of the storm witnessed by Humboldt in 1799 and of the European storm of 1832 drew attention, and many newspapers commented on the coincidence that all three meteor showers had occurred near the same date. Yale mathematics professor Denison Olmsted caught the last hour of the 1833 storm and analyzed observations from around the country. He established the shower's geographic extent and estimated that over 207,000 meteors were seen that night. Olmsted determined that the meteors had originated beyond earth and that they entered the atmosphere traveling in parallel paths. He concluded that the meteors were part of a nebulous body of unknown nature orbiting the sun and that the shower was caused by the earth's passage through this object.

Good Leonid displays occurred in 1835 (about 120 per hour) and 1836 (300 per hour), demonstrating that less activity might be expected each year. In 1837 the German physician and astronomer Heinrich Olbers looked at the available information and concluded that the main swarm of the Leonids returned every thirty-three or thirty-four years. "Perhaps we shall have to wait until 1867 before seeing this magnificent spectacle return," he wrote.

In the 1860s interest in the Leonids renewed with the historical research of Yale professor Hubert Newton, who combed Chinese, Arab, and European accounts of meteor showers looking for previous Leonid returns. He found records of eleven great meteor showers dating from A.D. 902, determined that the interval between Leonid storms was 33.25 years, and predicted the next return on the night of November 13-14, 1866. After a "warm-up" shower of one meteor a minute in 1865, a small Leonid blizzard returned on schedule, delighting observers from England to India with hourly rates over two thousand. Careful counts enabled astronomers to determine the exact moment of peak activity. The shower remained above half its peak value for sixty-five minutes, which indicated that the swarm of particles responsible for the performance spread across some 21,000 miles (33,800 kilometers) of space. Another fine display occurred the next year, with rates around a thousand per hour despite interference from moonlight, followed in 1868 and 1869 by progressively weaker encores.

The 1860s also brought a solution to the mystery surrounding the origin of the Leonids and the other meteor showers then being recognized. Orbital calculations of the Perseid meteor stream by Giovanni Schiaparelli, better known for his later work on Mars, revealed a very strong resemblance to the orbit of a bright comet discovered in 1862, and he argued that all meteor showers were caused by the disintegration of comets. The idea was quickly accepted by additional associations between comet orbits and meteor streams. Using information from the 1866 Leonid shower, Urbain Le Verrier published its orbital characteristics in 1867 — and immediately astronomers pointed out the similarity between the Leonid orbit and that of newly discovered Comet Tempel-Tuttle.

Comets are now considered "dirty snowballs" containing a mixture of dust and frozen gases. They only become visible near their closest approach to the sun, where areas on the comet's icy surface can become warm enough to evaporate. The resulting jets of evaporating gases carry with them any solid matter mixed in with the original ice (Fig. 7-3). At first concentrated around the comet, in time the debris gradually diffuses along the comet's entire track. A meteor shower occurs on the date in the year when the earth passes nearest to a comet's dust-littered orbit. The clumpiness evident in the Leonid swarm is an indication of its youth — the clouds of

Table 7-1. Major Meteor Showers. Dates of maximum activity vary from year to year; check the Appendix for the correct date in any given year. Hourly rates represent the average number of meteors visible for an observer under a clear, moonless sky; rates for some showers also fluctuate from year to year.

Shower	Date of maximum (UT)	Days above ¼ strength	Hourly rate	Appearance	Source body
Quadrantids	Jan. 3	< 1	40-100+	Average mag. +2.8; blue; medium speed	————
Lyrids	Apr. 21	2	10-15	Average mag. +2.4; swift	Comet Thatcher (1861 I)
Eta Aquarids	May 4	3	20	Average mag. +2.9; often yellowish; very swift	Comet Halley
Southern Delta Aquarids	July 29	7	15-20	Average mag. +3.0; often yellowish; medium speed	————
Perseids	Aug. 12	5	50	Average mag. +2.3; white, yellow, orange, and other colors; swift	Comet Swift-Tuttle (1862 III)
Orionids	Oct. 22	2	25	Average mag. +3.1; very swift	Comet Halley
Southern Taurids	Nov. 3-5	Broad maximum	< 15	Average mag. +2.8; slow	Comet Encke
Leonids	Nov. 17	4	10-15	Average mag. +3.0; very swift	Comet Tempel-Tuttle
Geminids	Dec. 14	3	50-80+	Average mag. +2.5; yellowish; medium speed	Asteroid 3200 Phaethon

particles simply haven't had time to become uniformly distributed along Tempel-Tuttle's orbit.

This new understanding of the origin of meteor showers, coupled with the terrific track record of the Leonids, gave astronomers confidence that another meteor storm would fill the November skies of 1899. Newspapers in Europe and America spread the word. But despite a promising splash of a hundred meteors an hour in 1898, a spectacular shower failed to appear — "the worst blow ever suffered by astronomy in the eyes of the public," wrote one astronomer. As later investigations would show, close encounters with Jupiter and Saturn after 1869 slightly altered the course of the Leonid swarm, causing them to pass well inside the earth's orbit. What peak there was finally occurred in 1901 — two years late — with a spurt of activity reaching 400 per hour as seen from the southwestern United States. And the next time around, between 1931 and 1933, the Leonids barely did

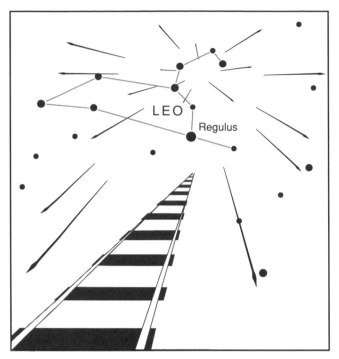

Figure 7-2. *Parallel railway tracks converge to a vanishing point in the distance. If you backtracked the paths of the meteors you see on November 17, most will converge on a point within the constellation Leo — the radiant of the Leonid meteor shower. The particles that make these metors travel together in nearly parallel paths.*

conditions are met in 1998 and 1999 — but they were also met in 1902 and 1933, years in which the Leonids produced no great activity. The possibility of an unusual shower event in 1998 or 1999 is very good, Yeomans notes, but it is by no means certain. "That's the way it is with meteor showers," he says. "You can say 'probably,' but if you say 'definitely' they'll get you every time."

No one really knows what to expect from the ever-erratic Leonids in the 1990s. The International Meteor Organization, which coordinates the activities of a global network of amateur observers, has already begun its International Leonid Watch and will monitor the shower through 1999. As the shower's history makes clear, impressive displays may occur in the years before or after the expected storm. Even if a true meteor storm fails to materialize, the Leonids still could produce a shower with rates higher than any of the other annual streams. The best bet for casual observers: Make it a point to check out the Leonids every year. We'll discuss the circumstances for the 1998 and 1999 Leonids in the next section.

better than half that count. It looked as though the once-great Leonids had finally begun to peter out.

Astronomers held out little hope for a meteor storm in 1966, despite indications of enhanced activity after 1963. Judging by the 1899 and 1901 returns, a close approach to the Leonid stream no longer seemed very likely, although some astronomers predicted that activity might reach hourly rates around one hundred. In fact that was a fair estimate of what most of the world witnessed on the morning of November 17, 1966. But die-hard amateurs in the western United States were treated to the storm to beat all Leonid storms! At 2:30 A.M. the count was 33 per hour, but by 4:00 it had increased to almost 200 and continued to rise. By 5:00 Leonids poured out of the sky at the rate of 30 per minute, and near 6:00 the count peaked at an incredible 40 meteors per *second!* The 1966 Leonid storm was one of the wildest displays in history — 144,000 meteors per hour at maximum.

So what's in store for us on the next return of the Leonid swarm? According to a 1981 study of Leonid behavior by astronomer Donald Yeomans, significant showers are only possible about six years before or after Comet Tempel-Tuttle reaches perihelion — between 1992 and 2004, this time around. Our best chance for a major shower event occurs when the earth brushes past Tempel-Tuttle's orbit shortly after the comet passes through. These

Figure 7-3. *The dust-laden jets of Halley's comet were viewed close-up by the Giotto spacecraft in 1986. This picture, a composite of sixty separate images, shows bright jets of gas and dust streaming away from the comet's sunward side. Debris from Halley creates two of the year's best meteor showers — the Eta Aquarids of May and the Orionids of October. Courtesy of Harold Reitsema and Alan Delamere, Ball Aerospace, and Max Planck Institut für Aeronomie.*

Observing Meteor Showers

A first-time meteor watcher, age nine, described his experience to us as "kind of like fishing" — and that's a pretty fair description. The best way to enjoy a meteor shower is to dress warmly, set down a blanket or lawn chair at a dark site, get comfortable, and watch the stars. There's no need to look directly at the radiant — meteors will appear in virtually every part of the sky. Naturally, interference from streetlights or the moon will reduce your count.

On any night of the year, meteors appear faster, brighter, and more numerous after midnight. That's when your location has turned into the earth's direction of motion around the sun and plows into meteor particles nearly head-on, rather than having them catch up from behind. The peak activity of a meteor shower occurs in the hours when the earth passes closest to the orbit of the shower particles. The ideal circumstance for any observer is for the shower to peak at a time when its radiant is highest in the sky during the morning hours, and most of the year's best showers have the potential to meet these criteria. Table 7-1 lists information on the major meteor showers, including the approximate date of maximum activity, typical hourly rates at peak, and the general appearance of the meteors. The calendar date for a given shower's peak varies a day or so due to leap years; refer to the Appendix for the correct date in a given year. Most meteor streams are spread out enough that looking at the actual hour of peak activity will make little difference in what you see. That's not true of the Geminids, Quadrantids, and of course the Leonids, whose streams remain quite clumpy; for these showers the appendix also lists the estimated hour of their peaks.

Most of the meteors you'll see during a shower are caused by fluffy particles not much larger than sand grains. As the particle enters the atmosphere it collides with gas atoms and molecules and becomes wrapped in a glowing sheath of heated air and vaporized material boiled off its own surface. Meteors become visible at altitudes between 50 and 75 miles (80 and 120 kilometers), with the faster particles typically shining at greater heights. Many of the faster, brighter meteors may leave behind *trains*, a dimly glowing trail that persists for many seconds, or more rarely, minutes. Larger debris may create a *fireball*, a spectacular meteor bright enough to outshine even Venus. Occasionally a fireball may break up, an event accompanied by bright flares and even "sparks" thrown a short distance from the meteor's main trail. Such a fireball is called a *bolide*.

Although we've discussed the Leonids at length, we've neglected the other showers listed in Table 7-1. The following notes will get you more acquainted with these meteor showers.

QUADRANTIDS

Generally visible between December 28 and January 6, the Quadrantids have a sharp activity peak around January 3. As few as forty and as many as two hundred Quadrantids have been seen during the shower's maximum, so although the stream is compact it's also apparently clumpy. Typical rates vary between forty and a hundred per hour and about 5 percent leave trains. The speed of the meteors, 26 miles (42 kilometers) per second, is moderate because the stream intersects the earth's orbit quite steeply (Fig. 7-4). When the shower was first recognized as annual in 1839, the radiant occurred in a constellation now no longer recognized — Quadrans Muralis ("Wall Quadrant"). It's now divided between Hercules, Bootes, and Draco. The cold nights of northern winters and typically faint meteors keep the shower from being truly popular.

LYRIDS

The Lyrids appear from April 16 to 25 and peak around April 21 (10-15 per hour); the radiant lies between Hercules and Lyra. Although it was not recognized as an annual shower until 1839, Chinese observations of the Lyrids date back to 687 B.C. — making it the earliest recorded meteor shower. Despite the low annual rate, the Lyrids have the capacity for impressive displays of over fifty per hour. This happened in 1803, 1922, and 1982, weakly suggesting a sixty-year period for Lyrid outbursts. In 1867 the shower was linked to its parent comet (Thatcher, 1861 I). The Lyrids are bright, rather fast (30 miles or 48 kilometers per second), and about 15 percent leave persistent trains.

ETA AQUARIDS

The first of the year's two showers derived from Halley's comet, the Eta Aquarids occur from April 21 to May 12 with a peak (10-20 per hour) around May 4. This shower is best for southern hemisphere observers, where the hourly rate climbs to fifty. The radiant is located near the Y-shaped asterism in Aquarius and named for one of those stars. The shower was discovered in 1870 and linked to Halley in 1868. The meteors are among the fastest (42 miles or 67 kilometers a second), faint on average, but the brighter ones have a yellowish color; about 30 percent leave trains.

SOUTHERN DELTA AQUARIDS

This is the most active stream of a diffuse group of streams and, as the name suggests, is best for the southern hemisphere. They may be seen between July 14 and August 18 and peak near July 29 (15-20 per hour). The meteors are medium speed (27 miles or 43 kilometers per second), tend to be faint, and few leave trains.

PERSEIDS

The best known of all meteor showers, the Perseids never fail to put on a good show and, thanks to its late-summer peak, are usually widely observed. The earliest record comes from China in A.D. 36. Generally visible from July 23 to August 22, the speed (37 miles or 60 kilometers per second), brightness, and high proportion of trains (45 percent) distinguish the Perseids from other showers active at this time. (See Fig. 7-4.) The shower peaks around August 12 with observed rates usually around fifty, but ranging to over a hundred per hour. Average hourly rates climbed from sixty-five between 1966 and 1975 to ninety between 1976 and 1983. The first meteor shower linked to a comet, the Perseids derive from Comet Swift-Tuttle. The comet was last seen in 1862, and had been expected to return in the early 1980s, but it didn't show up. Faint hope was held out for a 1992 return, based on the possibility that a comet seen in 1737 was also Swift-Tuttle. When the 1991 and 1992 Perseids showed unusual activity, expectations for a 1992 return were heightened, and in fact Swift-

Tuttle did appear in late 1992. The comet was visible in northern skies in the fall and early winter. At this writing, it's not known what the effect of the "lost" comet's return will have on the Perseids in 1993 and beyond — but you can see for yourself!

ORIONIDS

This is the sister stream of the Eta Aquarids, both of which arise from the debris of Halley's comet. Orionid meteors can be found between October 15 and 29, with a peak of about twenty-five per hour around October 21. Discovered in 1864, the Orionids were not linked to Halley until 1911. Orionid meteors are among the fastest (42 miles or 67 kilometers per second), generally faint, and about 20 percent leave trains that persist one or two seconds.

SOUTHERN TAURIDS

Visible between September 15 and December 15, this is the strongest of several streams originating from Encke's comet. A broad maximum occurs between November 3 and 5, but usually brings an hourly rate less than fifteen. The shower was first recognized in 1869 and associated with Encke in 1940. Its meteors are generally faint and quite slow (19 miles or 30 kilometers a second) because they approach the earth from behind and must catch up.

LEONIDS

These are generally seen from November 14 to 20, with a peak hourly rate on November 17 of between ten and fifteen per hour; about half leave trains that can persist for several minutes The shower's most notable feature, of course, are its periodic outbursts related to Comet Tempel-Tuttle, which comes to perihelion again in February 1998. No one knows if the Leonids will cooperate, but the moon

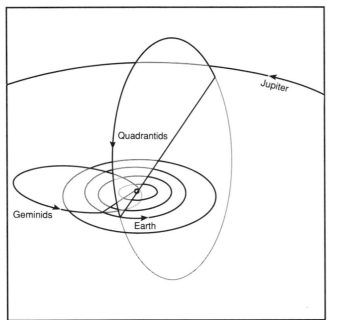

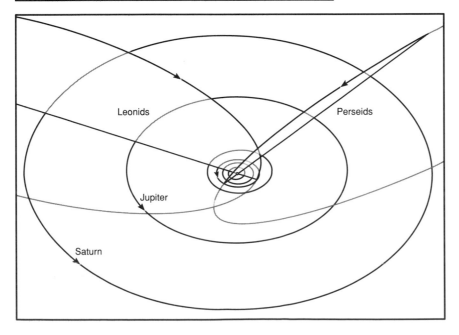

Figure 7-4. *Top: Particles in the Geminid meteor stream hit the earth near the midnight side, making it one of the few showers best seen in the late evening. Similarly, the Quadrantid meteors plunge steeply downward into the earth's orbit. The meteors of both showers are of moderate speed because they're neither catching up to the earth nor being plowed into by it. Bottom: Earth runs nearly head-on into particles of the Perseid and Leonid swarms, making their meteors among the swiftest.*

is just two days from new for the 1998 shower and effectively out of the way. The moon is not quite so cooperative for the 1999 shower. It's two days past first quarter, which means that it's low in the west as the radiant rises — less than ideal, but not really bad. We strongly recommend that you make a point of observing the Leonids prior to these years as well. We have attempted to pick the likeliest time for the storm's appearance, but as we've seen, Leonid predicting is a very uncertain business. The authors plan to be out all night in 1998 and 1999; all we can really predict is that only those who are out there looking when it happens will see it!

Because the earth runs into the orbiting particles almost directly head-on, Leonid meteors travel faster than those of any other shower — 45 miles (72 kilometers) per second (Fig. 7-4). It's unclear what effect the recurrence of a Leonid storm might have on satellites and spacecraft in Earth orbit. Although there were no reports of satellite problems caused by the 1966 Leonids, the number of a satellites has greatly increased since then. "The real storms from the Leonids are extremely concentrated and usually rich in small, young particles," says Peter Brown of the International Meteor Organization. The very smallest particles are stopped in the thin upper atmosphere and cause no flash. "So the number of visual meteors observed on the ground," Brown notes, "may actually be a small fraction of the total number of particles present." At Leonid speeds, even the smallest particles can pack quite a punch.

GEMINIDS

The Geminids are active between December 6 and 19 and peak near December 13, with typical hourly rates around fifty but ranging to eighty and higher. Always second to the August Perseids, the Geminids began improving in the 1960s and now hold the title of "year's best meteor shower." Because the Geminids intersect Earth's orbit near the midnight side, this shower is one of few that are quite good before midnight (Fig. 7-4). Its meteors are medium speed (22 miles or 36 kilometers per second), bright (13 percent magnitude 0 or brighter), and appear white or yellowish. The parent body of the Geminids, discovered only in 1983, makes it so far unique among meteor showers. The body, named 3200 Phaethon, is classed as an asteroid, one of thousands of rocky objects whose orbits generally keep them between Mars and Jupiter. Eventually all the icy substance of a comet will evaporate, perhaps leaving behind a rocky core that forms an "extinct comet" class of asteroids. Many scientists suggest that perhaps as many as one-third of the asteroids whose orbits cross that of Earth are really defunct comets!

Rocks from the Sky

As we've already noted, most of the meteors you're likely to see arise from debris no larger than pebbles. For example, a one-ounce (33 gram) pebble traveling at the cosmically slow pace of 19 miles (30 kilometers) a second will create a meteor bright enough to outshine all first-magnitude stars (Fig. 7-5). Yet even objects this small may not be completely destroyed by their fiery passage through our atmosphere, and some will reach the earth's surface as *meteorites*. Astronomers estimate that each day the earth sweeps up more than five hundred objects weighing at least three ounces (100 grams) — about three hundred fifty of which fall onto an area the size of the United States every year!

Vague stories about rocks falling from the sky occur in many ancient writings, but it's clear that at least some peoples held meteorites in high regard. The iron Casa Grande meteorite was found wrapped in linen and buried under the floor of a Montezuma Indian temple in Chihuahua, Mexico. The sacred black stone built into the Ka'ba at Mecca, the chief object of Muslim pilgrimages, is also likely a meteorite.

Iron meteorites, easily distinguished from terrestrial rocks, played a role in the development of tool-making in some areas. In China, the iron blade of an axe found in a Shang dynasty (ca. 1,400 B.C.) tomb was clearly forged from a meteorite. Polar explorers in the nineteenth century found that West Greenland Eskimos used a trio of meteorites for the nickel-iron tips of their weapons and tools. One of these irons, the fifty-nine-ton *Ahnighito* ("the Tent"), is the largest ever displayed in a museum and can be seen in New York City's American Museum of Natural History. Only one larger meteorite has so far been found — the sixty-six-ton Hoba iron, which still rests where it fell in Namibia.

Modern science resisted any extraterrestrial explanation of meteorites until proof literally fell from the sky. At midday on April 26, 1803, before hundreds of witnesses, a meteorite flew over the French town of L'Aigle and exploded in a shower of three thousand stones. When small stones fell on Weston, Connecticut, in 1807, investigators from Yale University concluded that they had fallen from the sky. President Thomas Jefferson was openly critical: "I would sooner believe that those two Yankee professors would lie than believe that stones would fall from heaven," he said.

About three thousand meteorites have so far been recovered, and since none bear any resemblance to the fragile, low-density material from which comets and their meteor showers are made, meteor showers present no greater chance of a meteorite fall. Most meteorites are believed to be chips broken off asteroids during collisions. In the early 1980s one unusual meteorite found in Antarctica bore an astonishing resemblance to igneous rocks returned by the Apollo missions to the moon. Blasted into space by the blow of some great meteorite impact, this rock is now accepted as a meteorite from the moon. Even more incredible, several other intriguing meteorites contained gases in ratios strikingly similar to those of the

Figure 7-5. *A meteor rushes through the sky near the dusty tail of Comet West in March 1976. Photo by Geoff Chester, Albert Einstein Planetarium.*

martian atmosphere and probably originated from the surface of Mars.

With all of this material falling from the sky, it would seem that sooner or later someone could be hurt. One estimate of meteor impacts on humans determined that there should be one victim every 180 years in an area the size of North America. But amazingly one documented injury from a meteorite *is* recorded. On November 30, 1954, a meteorite weighing in at 8.6 pounds (3.9 kilograms) penetrated the roof and ceiling of a house in Sylacauga, Alabama, bounced off a large radio and inflicted painful bruises to a woman asleep on a nearby couch! Extrapolated to the entire world, these estimates would suggest an average of one meteorite-related injury every nine years, although few will be as dramatic — or painful — as the Sylacauga event.

Meteorites far larger than any recorded have struck the earth in the past, and some may have altered the progress of life on our planet. Mounting evidence suggests that at least one large impact 65 million years ago, possibly at a site along Mexico's Caribbean coast, may have been the final curtain for the dinosaurs and many other species, but a tremendous ecological opportunity for mammals. Scars from such impacts are quickly eroded on earth, yet about one hundred thirty craters are known. Astronomers are now attempting to survey all objects in earth-crossing orbits in order to locate any that may prove a future threat.

In ancient China meteors were viewed as messengers from heaven, their speed a measure of the importance of the message. Apart from the lunar samples returned during the 1960s and 1970s, meteorites represent the only samples of extraterrestrial material available for direct study. Their messages, eagerly deciphered by planetary scientists, have revealed volumes about the history of the earth and the origin of the solar system.

8   UNPREDICTABLE EVENTS

Throughout this book we've emphasized the motions of the moon and planets, listed the best opportunities for seeing them, and pointed out some of their more interesting gatherings. This may give the impression that every sky event worth viewing has already been predicted. Fortunately for those of us who look forward to surprises, there are three types of naked-eye sky events *no one* can predict yet. We discuss them below in order of the likelihood of their appearance by 2001.

The Northern Lights

The aurora borealis is a sporadic, generally faint, atmospheric phenomenon usually seen in the night sky from locations at high latitudes. More commonly known as the "northern lights," it may first appear as a faint milky glow low in the north, too dim for the human eye to detect any color but bright enough to silhouette any clouds near the horizon. From there it may develop into steady greenish arcs or form scintillating, swirling curtains of yellow-green light. During the most dramatic displays visible from middle latitudes, a crimson glow fills much of the sky. It was this form that inspired European scientists of the 1600s to name the phenomenon *aurora borealis*, literally "northern dawn." (The corresponding phenomenon in the southern hemisphere is called *aurora australis.*) The patterns and forms of the aurora include quiescent arcs, rapidly moving rays and curtains, patches, and veils. Since this light show takes place high in the atmosphere, it isn't really an astronomical event. But, as we'll see, the aurora represents the visible manifestation of complex interactions between earth and the sun.

The lively, colorful forms of the aurora naturally appear in much American and European folklore. A Scottish legend connects the lights with supernatural creatures called Merry Dancers, who fight in the sky for the favor of a beautiful woman. A Danish story explains the lights as reflections from flocks of geese trapped in the northern icepack and flapping to free themselves. The Lapps in Sweden believed that the aurora could scorch the hair of those foolish enough to leave the house without a cap. In Alaska and eastern Norway, children were believed to be at special risk from the northern lights. Greenland Eskimos thought the aurora represented signals from dead friends who were trying to contact the living. The Fox Indians of Michigan believed that they could conjure up spirits by whistling to the light. To the Tlingit Indians of southeastern Alaska, the flickering glow was caused by battles between the spirits of fallen warriors; its appearance foretold catastrophes and bloodshed.

There are many historical accounts of the northern lights from areas far south of its usual location, such as southern Europe. The Greek philosopher Aristotle is often credited as the first in Western culture to attempt to discuss the aurora scientifically. He gives several names for the northern lights in his *Meteorologica*, written in the fourth century B.C. These include "burning flames," "chasms," "trenches" — and "goats," apparently an allusion to a very active form.

The predominant color of bright auroras seen at the latitudes of the Mediterranean — and the continental United States — is a deep red. An early Chinese record describes it as a "red cloud spreading all over the sky." According to the Roman philosopher Seneca, in A.D. 37 troops moved during an aurora to assist the seaport of Ostia "as if it were in flames, when the glowing of the sky lasted through a great part of the night, shining dimly like a vast and smoking fire." Similar "fires in the air" in 1583 mobilized thousands of French pilgrims, who prayed to avert the wrath of God. On September 15, 1839, an intense aurora dispatched fire departments throughout London. More recently, during the great auroral display of March

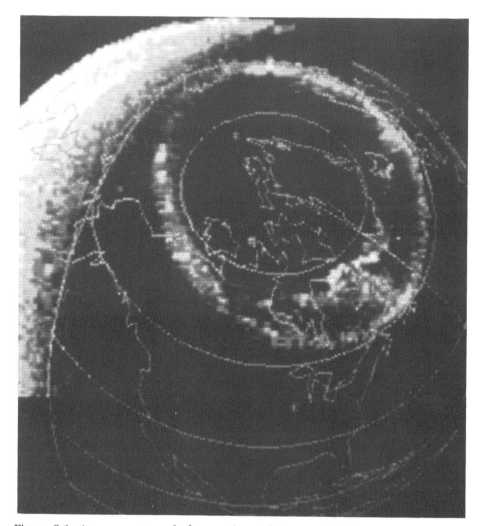

Figure 8-1. *A permanent oval of auroral activity crowns both the earth's polar regions. A satellite-borne ultraviolet camera caught the northern oval on November 8, 1981, when it expanded to the U.S.-Canadian border. Photo by NASA*

13, 1989, police departments and newspapers across the United States fielded telephone calls about "funny red clouds" and a fire-like glow in the sky.

The aurora occurs in two great luminous ovals centered on the earth's north and south magnetic poles (Fig. 8-1). Their glow results from collisions between atmospheric gases and showers of electrons guided by the earth's magnetic field. Each gas gives out its own particular color when bombarded, and atmospheric composition varies with altitude. The auroral glow can originate 50 to 621 miles (80 to 1,000 kilometers) above us, but typically occurs between 62 and 155 miles (100 to 250 kilometers) above the ground. Oxygen atoms provide the yellow-green color most commonly seen in auroras, while molecules of nitrogen and molecular oxygen at lower altitudes provide blue and red light. These three primary colors together produce the myriad hues of a typical aurora.

What causes the showers of electrons that create the northern lights? Ultimately the source lies in the *solar wind*, a fast-moving stream of particles constantly flowing from the sun that carries the sun's magnetic field out into space. The solar wind molds the earth's magnetic field into an elongated bubble or cavity, compressing its sunward side and stretching its night side far beyond the moon's orbit. Under certain conditions the solar wind magnetic field can merge with the earth's, creating electrical currents that drive electrons into the polar atmosphere.

The solar wind also has its equivalents of gusts and gales. Transient events occurring on the sun can generate shock waves that greatly intensify the solar wind's impact on the earth. Solar flares may eject material from the sun's surface for hours. Areas called coronal holes generate broad torrents of solar wind and may last for many months. When the shock waves produced by these events slam into the earth's magnetic field, they may accelerate clouds of particles toward earth. This blast of energetic particles can transform a quiet aurora into a spectacular display. The auroral ovals expand greatly, bringing the northern lights to skygazers at far lower latitudes than normal. Impressive displays occurred in March, June, and November 1991.

Both solar flares and coronal holes are more common around the active portion of the sun's eleven-year sunspot cycle. This peaked in mid-1989, so solar activity is now on the downswing and will continue to decline until 1996, when the next cycle begins and activity slowly rises. Major flares should begin reappearing by 1999 in advance of the solar activity maximum expected in 2000. According to Cary Oler, Operations Manager of the Solar Terrestrial Dispatch in Stirling, Alberta, Canada, observers in the northern United States (north of forty degrees latitude) may expect to see between one and five auroras every few months until 1995. Observers at lower latitudes will continue to see displays associated with flaring solar regions or coronal holes, but with decreasing frequency — perhaps one or two each year. Auroral activity also intensifies in the spring and fall, largely because the earth's magnetic field is then more favorably oriented for coupling with the solar wind.

Overall, the chances of seeing an aurora are not all that bad — especially in Canada and the United States. Since the north magnetic pole lies in North America, the auroral oval generally reaches farther south there. This means that observers at a given latitude in North America have a better chance of seeing an aurora than those at the same latitude in Europe or Asia. Both Rome and Chicago lie at forty-two degrees north latitude, for example, but Rome averages one aurora per decade while Chicago sees about ten each year. Figure 8-2 shows the frequency of auroral displays in the northern hemisphere averaged over many years; the number in each curve indicates the number of nights per year in which an aurora can be seen.

Perhaps the strangest question connected with the northern lights is whether or not it produces sounds that humans can hear. Reports of rustling, hissing, and crackling noises associated with strong auroral displays have persisted for centuries. Sounds produced in the rarefied air in which auroras occur simply cannot reach the ground. Assuming that these sounds are indeed produced by the aurora — and not rustling leaves, insects, or active imaginations — one possible explanation for them proposes that, under certain circumstances, the aurora creates an intense electrical field that causes small discharges from ground objects. "There *may* be some credence to theories explaining the phenomenon through electrostatics," says Cary Oler. "My personal opinion is that the phenomenon doesn't exist, but I can't dismiss it as impossible since it may be observed under very special conditions." So when observing an active aurora, look *and* listen!

The atmospheric activity responsible for the northern lights occasionally has a profound effect on everyday life. "During the aurora of September 2, 1859," wrote the American researcher Elias Loomis, "the currents of electricity on the telegraph wires were so steady and powerful that, on several lines, the operators succeeded in using them for telegraphic purposes as a substitute for the battery." For a time, messages were transmitted solely on currents generated by the aurora. A rapidly shifting and expanding auroral oval can induce electrical currents in other long conductors as well. A dramatic example occurred in March 1989, when an extremely active solar region erupted with flares that broke records held for over thirty years. Auroral activity was seen as far south as Jamaica. In Quebec, Canada, induced currents saturated transformers, leading to a total collapse of the power distribution system that for nine hours left over six million people without electricity. Oil pipelines and telecommunications cables also experienced damaging electrical surges. Keep this in mind when an aurora next paints the sky.

Bright Comets

In the previous chapter we discussed comets only in the context of their relationship with meteor showers, but comets can be impressive sights on their own. In recent years, amateur and professional astronomers together have discovered an average of more than fifteen comets annually. Most of these remain diffuse fuzz balls detectable only in telescopes or, at best, binoculars. A truly bright comet, one that may be easily seen by the unaided eye, appears only about once each decade — and by the reckoning of many comet-watchers, we're long overdue.

Comets are dark, solid bodies a few kilometers across that orbit the sun in very eccentric paths. Comets can be fairly described as being "dirty snowballs" containing a mixture of dust and frozen gases. Some of the icy material — perhaps less than one percent — evaporates as a comet nears the sun, creating an envelope of gas and dust that enshrouds the solid body. This envelope, called the *coma*, may be up to one million kilometers (about 620,000 miles) across. Swept back by the solar wind and the radiation pressure of sunlight, this material forms the comet's tail. Comet tails often span tens of millions of kilometers and sometimes reach lengths greater than the distance between earth and the sun. That such a small amount of material creates visible features so large has led some to describe comets as "the closest thing to nothing anything can be and still be something."

To the eye the coma of a bright comet looks almost star-like, a tiny ball of light set within a milky glow. The comet's tail or tails fan out from the coma. If present, a broad dust tail may be the most striking visual feature, arcing across ten degrees of sky or more. The glowing gas tail is straighter, narrower, and fainter than the dust tail. Within the coma, and invisible to both the naked-eye and the most powerful telescopes, lies the small icy body responsible for this grand apparition — the comet's *nucleus*.

The ancient Chinese names for comets reflect their visual appearance. A comet with a prominent tail was called a "broom star" (*hui-hsing*), while one with no

Figure 8-2. *This map, centered on the north magnetic pole, gives an idea of the average frequency of auroral displays for different locations in the northern hemisphere. Numbers correspond to the number of nights per year in which an aurora can be seen. The shaded region shows the zone where auroral displays are most common.*

obvious tail was a "bushy star" (*po-hsing*). (Our word "comet" comes from the Greek word for "long-haired one.") Until the mid-1400s the most detailed and complete observations of comets were made by the Chinese, who as early as 200 B.C. employed official skywatchers to record and interpret any new omens in the heavens. They recognized that comet tails always point away from the sun some nine centuries before their European counterparts. The Chinese interest in comets, however, was for their astrological importance as signs of coming change. In A.D. 712, after his astronomers reported the terrible implications of a recent bright comet, the T'ang dynasty emperor Jui-tsung elected to abdicate his throne rather than risk disaster.

Oriental ideas about comets had little influence on the development of Western thought. While some argued that comets were planet-like astronomical bodies, the influential Aristotle regarded them as a fiery atmospheric phenomenon. They could not be planets, he reasoned,

because comets can appear far from the constellations of the zodiac. Aristotle envisioned comets as being whipped up by the motion of the sun and stars around the earth, and their appearance warned of coming droughts and high winds. The Roman philosopher Seneca and the Egyptian astronomer Ptolemy both discussed the use of comets in weather forecasting. As these ideas were extended in the Middle Ages, comets became less a portent of disaster and more a cause. They were viewed as a fiery corruption of the air, pockets of hot contaminated vapor that could bring earthquakes, disease, and famine.

Some of these ideas were being seriously questioned when the great comet of 1577 attracted the attention of Danish observer Tycho Brahe, recently installed in his observatory on the island of Ven near Copenhagen. Tycho could see no reason why comet tails should always point away from the sun if they were products of the weather. He measured the position of the comet with respect to the stars at different times during the night in an effort to find

Table 8-1. Notable Comets. Roman numeral designations, names, perihelion distance (as percentage of the earth-sun distance), number of years until next return, and visual rankings of recent comets.

1965	VIII	Ikeya-Seki	0.8%	880	Very good
1970	II	Bennett	53.8%	1,700	Excellent
1973	XII	Kohoutek	14.2%	80,000	Poor
1976	VI	West	19.7%	6,400,000	Excellent
1983	VII	IRAS-Araki-Alcock	99.1%	1,000	Good
1986	III	Halley	58.7%	75	Poor

its parallax, a clue to the object's true distance from Earth, and these observations indicated that the comet lay beyond the moon but not as far off as Venus. Tycho's work on the comet of 1577 did not settle the matter — Galileo, for example, dismissed his observations — but it did help break the hold of Aristotle and hasten the scientific study of comets. When Isaac Newton published his monumental *Principia* in 1687, he showed that comets obeyed Kepler's laws of planetary motion and concluded that "comets are a sort of planet revolved in very eccentric orbits around the sun."

Future observations of the comet of 1682 would eventually remove any lingering doubts. Newton's friend Edmond Halley began collecting accurate cometary observations in 1695 with the goal of comparing the orbits of many comets. Looking over his table of orbits, Halley found several comets that seemed very similar and shared roughly the same period, between seventy-five and seventy-six years. "Many considerations incline me to believe the comet of 1531 observed by Apianus to have been the same as that described by Kepler . . . in 1607 and which I again observed in 1682," Halley wrote. "Whence I would venture confidently to predict its return, namely in the year 1758. And if this occurs, there will be no further cause for doubt that the other comets ought to return also." Halley's confidence proved well founded — the first comet ever predicted to return was again spotted on December 25, 1758, and has been known as Halley's comet ever since.

Comets are more commonly named for their discoverers; up to three independent co-discoverers may share the credit. Comets also receive letter designations by the year of discovery and order of discovery within the year — the first of 1993 would be 1993a, the next 1993b, and so on. Thus Halley's comet, 1982i, was the ninth comet discovered (or in this case *recovered*) in 1982. Once the orbit is computed, comets receive a more permanent designation with Roman numerals indicating the order of perihelion passage within each year. Halley, the third comet to make its closest approach to the sun in 1986, is 1986 III under this system.

As of this writing, astronomers have detailed orbital information on 841 individual comets. Of these, only 170 are periodic comets, those that complete an orbit in less than 200 years. Halley's comet is the brightest and most active member of this group. The remaining 671 are called long-period comets, which take 200 years or more to return to the inner solar system. According to Daniel Greene, an astronomer at the Smithsonian Astrophysical Observatory in Cambridge, Massachusetts, no comets are reliably predicted to become bright enough for a good visual display by 2001. So comet afficionados pin their hopes to the unanticipated arrival of an as yet unknown long-period comet.

Table 8-1 lists information on the best comet appearances since 1965, including a simple ranking of their visual displays. The two most important considerations in assessing the visibility of a comet are its distance from the sun at perihelion, which controls the comet's activity, and its distance from earth, preferably after the intense heating of perihelion. Comet West improved dramatically within a week of its very close approach to the sun, aided in part by the break-up of its nucleus into four fragments. West was the last "great comet," dominating the morning sky of early March 1976 with gas and dust tails up to twenty-five degrees long (Fig. 8-3). It was even bright enough to be seen from large cities.

Comet Halley was an impressive sight in 1910, but its anemic 1986 appearance was disappointing even for those who traveled far from city lights. The main difference between the apparitions was the comet's distance from earth. Halley reached perihelion at a time when the earth was on the opposite side of the sun, and the comet never came closer to earth than 39 million miles (63 million kilometers) — about three times the distance of its 1910 approach.

Another example of the importance of proximity was the 1983 display of comet IRAS-Araki-Alcock. A small and relatively inactive comet, it was discovered first by the Infrared Astronomical Satellite (IRAS) in late April and originally identified as an asteroid. In early May, amateurs Genichi Araki of Japan and George Alcock of England independently discovered the object. It soon became an obvious naked-eye sight high in the northern sky, and on May 12 the comet passed within 2.9 million miles (4.7 million kilometers) — closer to earth than any comet since

Figure 8-3. *Comet West, the last great comet, gave skygazers a grand show on the mornings of March 1976. The comet was bright enough to be seen near large cities, but was most impressive from dark skies. Photo by Ronald E. Royer and Steve Padilla.*

1770. A typical comet might move across the sky by a degree or so a day, too slowly for the eye to notice. But IRAS-Araki-Alcock was so close that its motion was clearly evident to observers, who compared its movement to that of the minute hand on a clock. At its best, the comet was about twice the apparent diameter of the moon and looked like a star nestled within a puff of smoke. It showed no evidence of a tail — a fine example of a "bushy star" — and faded from view by the third week of May.

Even when orbital geometry promises a good display, the comet itself may simply fail to cooperate. Comet Kohoutek, which was widely predicted to be the "comet of the century" in 1973, did manage to become a naked-eye object but never lived up to its publicity. A more recent example is Comet Austin (1990 V), discovered in December 1989 by New Zealand amateur Rodney Austin. The comet's orbit was very favorable, taking Austin 32 million miles (51 million kilometers) from the sun on April 9, 1990, and 23 million miles (37 million kilometers) from the earth on May 25. Astronomers noted in January that Austin was already more active than Halley at the same distance from the sun, and in February astronomical hobby magazines hopefully proclaimed "Monster Comet Coming!" But as Austin closed on the sun it failed to

maintain its rapid brightening and, in the end, proved a bigger dud than Kohoutek.

Both Austin and Kohoutek appear to have been new comets, comets making their first close pass by the sun. Astronomers believe that all comets originate in a vast swarm of trillions of comets that surrounds the solar system. This "comet cloud" begins at the edge of the planetary system, near the orbits of Neptune and Pluto, and extends one-third of a light-year or more into space. The feeble gravitational disturbances from passing stars and interstellar gas clouds is all it takes to remove orbital energy from some of these comets, forcing them on a million-year-long fall toward the sun. Dislodged comets may arrive from any direction, their elongated orbits randomly oriented to the orbits of the planets (Fig. 8-4). Repeated gravitational interactions with the planets may rob comets of additional orbital energy, forcing some into tighter orbits (the short-period comets). New arrivals from the comet cloud probably retain a coating of highly volatile ices, such as carbon dioxide, that begins to evaporate at much lower temperatures than frozen water. Such comets "turn on" at relatively large distances from the sun, but brighten only until the coating evaporates.

Comets are thought to be remnants of the cloud of dust and gas from which the sun and planets formed. In

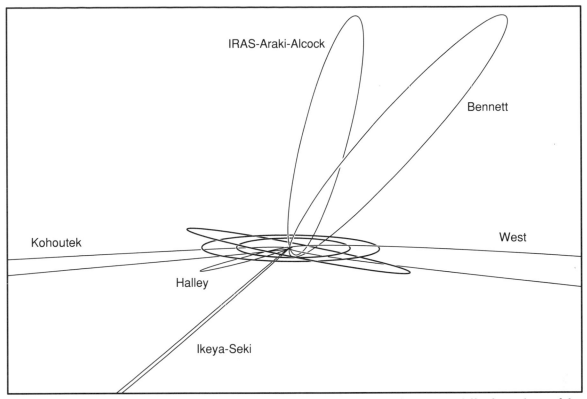

Figure 8-4. *This tangle of orbits shows how the eccentric paths followed by comets differ from those of the planets. All planetary orbits but those of Uranus, Neptune, and Pluto (heavy lines) have been omitted for clarity. This illustration shows the orbits of all the comets listed in Table 8-1.*

the deep-freeze of the outermost solar system, these fossils have remained largely unchanged in the four billion years since the birth of the solar system. For this reason, they are intensely studied by astronomers who see in them a chance to glimpse our most ancient past.

For the rest of us, a bright comet is simply one of the most spectacular sights in the celestial pageant. Astronomers reckon that a naked-eye comet appears roughly once every decade. The 1980s brought Comet IRAS-Araki-Alcock, certainly a bright comet, but nothing like Ikeya-Seki of the sixties or West of the seventies. Perhaps we're overdue for another spectacular comet, but only one prediction is safe: Sooner or later, the "monster comet" skygazers have hoped for will hover in the morning or evening twilight.

New Stars

On February 23, 1987, astronomer Ian Shelton at Las Campanas Observatory, Chile, became the first person in nearly four centuries to spot an exploding star with his own eyes. Located 160,000 light-years from us in the Large Magellanic Cloud, one of the small satellite galaxies that orbit our Milky Way, the star brightened slowly over eleven weeks and reached a visual magnitude of +2.9. At maximum brightness it radiated 200 million times as much energy as the sun. This stellar explosion, called a supernova, marked the cataclysmic death of a star. Astronomers have recorded more than six hundred supernovas in distant galaxies, but the event dubbed SN 1987a — the first supernova, bright or faint, to occur that year — was by far the brightest seen in modern times.

Astronomers had long known that some otherwise faint stars suddenly flare up, usually brightening by thousands of times and then, over a period of weeks, fading back into obscurity. Such a star was termed a nova, short for the Latin *nova stella* ("new star"). Ordinary novas are now known to occur in close binary star systems containing a normal star and a white dwarf, the compact, aged remnant of a star like the sun. Hydrogen-rich gas from the normal star streams onto the white dwarf's surface, where it accumulates and eventually explodes. Although the outburst appears quite violent, in fact only the surface material is thrown into space and the white dwarf remains intact. In 1934 astronomers Fritz Zwicky and Walter Baade realized that some novas were far more energetic than others — powerful enough to disrupt an entire star — and argued convincingly that these "super novas" were a distinctly different phenomenon. Type I supernovas occur in close binary systems where so much matter accumulates on a white dwarf that the star collapses and explodes. Type II events, such as 1987a, occur in stars much more massive than the sun when they collapse after exhausting the fuel supply in their cores.

Supernovas represent one of nature's greatest spectacles. Most of the chemical elements heavier than helium, including the iron and carbon in our bodies and the oxygen we breathe, were forged inside supernovas and dispersed into the galaxy long before our solar system was born. The blast wave of an exploding star may have even initiated the process that gave birth to our own solar system.

Supernovas are rare. In a spiral galaxy such as ours, containing about 100 billion stars, astronomers expect only three or four supernovas to explode each century. But we witness far fewer than this, since the galaxy is filled with clouds of dust and gas that obscure our view. We usually notice only those supernovas occurring within a few thousand light-years of earth. SN 1987a, positioned in a part of the sky with very little obscuring dust, was an exception.

Despite its importance to astronomers, 1987a was hardly an impressive sight. It could be viewed well only in the Southern Hemisphere, and even at peak brightness it remained fainter than the brightest stars in the Big Dipper. It was, after all, located in another galaxy. The last supernova in our own galaxy was seen in 1604, just a few years before the invention of the telescope. (Actually, since the star lay 33,000 light-years from Earth, the event really took place 33,000 years before human eyes saw it!) Surveys of reliable historical records have turned up at least seven earlier candidates. Table 8-2 summarizes information on all visible supernovas since A.D. 1000. No one knows when the next Milky Way supernova will blaze forth, but a look through historical descriptions of the past few explosions can give us an idea of just how impressive an event it could be.

Because of the intellectual climate in Europe before the Renaissance, the first supernova to attract wide attention appeared November 7, 1572. Many astronomers dismissed reports of the new star as idle talk — some waited two weeks before they even bothered to observe it, despite the fact that it was as bright as Venus! The chief reason for this was the Aristotelian belief in the immutability of the heavens, that nothing beyond the planets ever changed. Among the first Europeans to begin serious study of the star was Tycho Brahe, who because of poor weather did not notice it until his after-dinner walk on November 11. The star remained visible for fifteen months and did not show any movement, indicating that it was much farther than the planets. "Hence this new star is located neither in the region of the Element, below the moon, nor in the orbits of the seven wandering stars [including the sun and moon], but in the eighth sphere, among the other fixed stars," Tycho wrote. His famous protégé, Johannes Kepler, would also make detailed ob-

servations of a supernova — the last seen in the Milky Way galaxy — just thirty-two years later.

The imperial astronomers of China's Ming dynasty also noticed the new star, although their measurements of its position were far less accurate than Tycho's. "It was seen before sunset. At the time," they reported, "the Emperor noticed it from his palace. He was alarmed and at night he prayed in the open air on the Vermillion Steps." Their name for such an unusual star-like object was *k'o-hsing*, "guest star."

A supernova of similar brightness appeared on July 4, 1054. In 1928 the American astronomer Edwin Hubble correctly suggested that an expanding gas cloud called the Crab Nebula represented the debris of this explosion (Fig. 8-5). The supernova was first noticed in Constantinople, where it was associated with an outbreak of the plague. It's possible that Native American groups throughout western North America next recorded the explosion. More than a dozen rock carvings and paintings from that era show a crescent moon near a large star-like symbol. Observers in western North America would in fact see the supernova two degrees from the waning crescent moon just before sunrise. The moon would have moved several degrees by the time the Crab supernova was visible in China, which may explain why their reports make no mention of it. Chinese records indicate that the Crab supernova could be seen during daytime for twenty-three days — suggesting that it was at least as bright as Venus — and that it remained visible for twenty-one months.

A recent study of pottery made by the Mimbres people, who lived in New Mexico about a thousand years ago, lends support to the supernova interpretation of the Indian rock art. Geometric designs of astronomical significance appear on many of the earthenware bowls made by the Mimbres. Some markings appear to correspond to the number of days in the moon's sidereal and synodic periods. But one bowl features a stylized rabbit, a common moon symbol, clutching a circular object from which emerge twenty-three rays.

The best supernova on record occurred in 1006. First seen by astronomers in Egypt, it was discovered the next night in Japan, China, and in parts of Europe. A Baghdad observer wrote "Its rays on the earth were like the rays of the moon." In Egypt: "The sky was shining because of its light. The intensity of its light was a little more than a quarter of that of moonlight." From European latitudes the star skimmed the southern horizon, yet an observer in Switzerland wrote "A new star of unusual size appeared, glittering in aspect, and dazzling the eyes, causing alarm . . . It was seen likewise for three months in the inmost limits of the south, beyond all the constellations which are seen in the sky." The star's appearance also caused alarm in China, but an astronomer named Chou K'o-ming saved the day by deciding that the object was one of the

Table 8-2. Visible Supernovas Since A.D. 1000.

Year	Name	Constellation	Mag.	Distance (light-years)
1006		Lupus	-9.5	3,200
1054	Crab Nebula	Taurus	-4.0+	7,200
1181		Cassiopeia	0.0	26,000 or more
1572	Tycho's star	Cassiopeia	-4.0	20,000 or more
1604	Kepler's star	Ophiuchus	-3.0	33,000
1987	SN 1987a	Dorado	+2.9	160,000

"auspicious stars" and recommending that officials be allowed to celebrate.

These reports suggest that the supernova was extremely bright, comparable to a first-quarter moon. It's hard to imagine a star — a single point — that bright. Viewed through the unsteady air near the horizon, this brilliant star must have flickered madly and displayed rapid shifts of color. Chinese records indicate that the 1006 supernova remained visible for at least two years, although for observers in the far north it was seen for just a few months.

How will the next stellar outburst in our galaxy stack up to these historical events? Only time will tell, for the light of the next supernova is already heading our way . . .

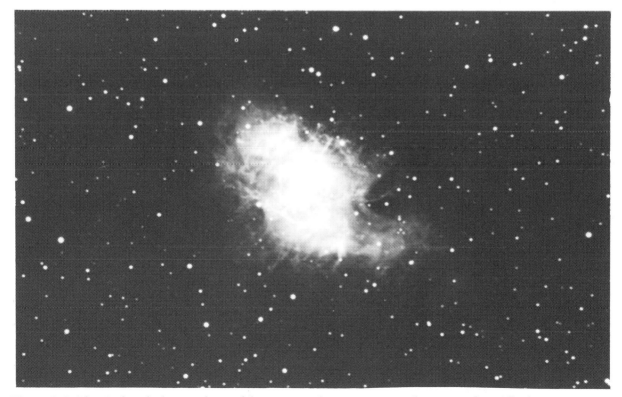

Figure 8-5. *The Crab Nebula, wreckage of the A.D. 1054 supernova, can be seen with small telescopes. Courtesy of California Institute of Technology and Carnegie Institute of Washington.*

CELESTIAL EVENTS, 1993–2001

The following pages contain an astronomical almanac for the years 1993 through 2001 — think of it as a "program guide" to the sky. It lists all important celestial happenings in the order of their occurrence and includes the moon's phases, lunar and solar eclipses, information on the brightness and visibility of the planets, and much more. Since this book focuses only on the most observable events, this is not intended to be an exhaustive list of planetary activity.

To find out what's in store for a given year, read the brief Sky Summary of events for that year. To find out what's happening on a given day, just turn to the nearest date in the almanac. For eclipses and other events that can be seen only over a limited area, we've listed the time as well. Events discussed or illustrated within the book contain a reference to the page number where you will find more detailed information.

Throughout the book we indicate the brightness of the planets and stars by referring to the standard astronomical magnitude scale. We usually give an object's magnitude parenthetically after the object's name — for example, Mars (+1.2) or Saturn (-0.3). Remember that the smaller the number, the brighter the object. Venus, the third brightest object in the sky, has a negative magnitude around -4.0, while stars at the limit of human visibility, magnitude +6.0, are 10,000 times fainter.

A Note on Time and Accuracy

We refer to time in several different ways throughout this book. Our horizon scenes illustrate the positions of the moon and planets as they'll be seen by observers in the United States and southern Canada at the times and dates noted in the captions. This is also true of notes in this almanac that instruct you to look at some time before or after sunset.

Our star maps illustrate the constellations as they appear at the times on your clock — your local time — even including the annual "spring ahead/fall back" changes of daylight savings time. If your area does not recognize these changes, then times on these charts between April and October are one hour ahead of the stars — so just look an hour earlier than indicated.

Some celestial events, such as eclipses or meteor showers, occur at specific times whether or not North America is angled to witness them. We list the occurrences of such events in Universal Time (UT), which is essentially Greenwich Mean Time (GMT), clock time on the meridian that runs through Greenwich, England. To convert UT to time in the United States and Canada, use the conversion table below. After you convert from UT to local time, remember that the time is still in a twenty-four hour format where 0h equals midnight, 12h is noon, and 18h is 6 P.M. And remember to keep track of the date! For example, observers in the Eastern time zone interested in an event that occurs at 4h UT on December 15 should expect the event at 23h or 11 P.M. Eastern Time on December 14. Everyone makes this mistake at least once!

One other caveat about time. While in many ways astronomy has served as the science of time, many events cannot be predicted with great precision. Also, most events visible to the naked eye don't *require* great precision — the unaided human eye simply isn't that precise an instrument. Times for some events, such as earth's arrival at perihelion or the peaks of some meteor showers, are quoted only to the nearest hour. For any special considerations regarding time, be sure to refer to the chapter in which the event is discussed.

Table A-1. Converting Universal Time.

To convert UT to:	For Standard Time, subtract:	For Daylight Time, subtract:
Atlantic	4 hours	3 hours
Eastern	5 hours	4 hours
Central	6 hours	5 hours
Mountain	7 hours	6 hours
Pacific	8 hours	7 hours
Most of Alaska	9 hours	8 hours
Hawaii	10 hours	———

Sky News

Magazines such as *Astronomy, Sky and Telescope, Natural History* and others are great sources for general sky information. There are also several electronic information sources that will keep you abreast of fast-breaking news, such as the next bright comet or predictions of auroral activity. All you need is a computer, a modem, and a telephone. There are many excellent electronic "bulletin boards" devoted to astronomical information; they are too numerous to mention here. These services are typically free, but often require long-distance phone calls. Two commercial services that offer up-to-date astronomical information are the General Electric Network for Information Exchange (GEnie) and the CompuServe Information Service (CIS). These nationwide systems require monthly fees; some surcharges may also be assessed. For most users, however, the connection only requires a local call.

For further information on GEnie call 1-800-638-9636. To learn more about CIS, call 1-800-848-8990.

For general astronomy news and discussions, CompuServe users may visit the electronic news service of *Sky and Telescope* magazine (SKYTEL) or the ASTROFORUM. GEnie subscribers can find astronomy news in Category 4 of the Space and Science Roundtable.

The Solar Terrestrial Dispatch in Stirling, Alberta, Canada, is a private research support center for institutions and organizations requiring timely solar and geophysical data on a regular basis. Its electronic bulletin board, open to all, contains detailed forecasts of solar and auroral activity — (403) 756-3008, 1200/2400 bps. These reports and maps are also distributed over many information services: CompuServe's HAMNET Forum, GEnie's Space and Science Roundtable (Category 2, Topic 11), various Internet news groups, and many others.

1993 Sky Summary

As the year begins, **Mars** and Venus are our main evening targets. Mars is at its best during the first two weeks of January; it drifts slowly toward the sun for the remainder of the year. **Venus** is brilliant in the west, peaking in its evening apparition in early February. It swings past the sun as spring begins, then assumes the role of morning star for the rest of the year.

Jupiter and **Saturn** lie nearly opposite each other in the sky. Saturn is visible low in the west in January, but quickly moves to the morning sky; it doesn't grace the evening sky until autumn begins. Jupiter, on the other hand, is a spring and summer planet. It begins the year high in the south at sunrise. Jupiter's shift to the evening sky comes as spring begins: the giant planet rises at sunset in late March and remains with us through the warm summer evenings. Toward the end of summer, Mars passes nearby.

Mercury puts on a fine show in 1993. Of its six apparitions, three will reward your attention with a fourth of borderline interest. Mercury visits Venus in the evening during February and again in the morning in August. It joins Jupiter and Venus in the morning sky in November. A special event is the planet's transit across the face of the sun in November. (You'll need a telescope and safe observing technique to enjoy this rare sight. See Chapter 3 for tips on viewing the sun safely.)

The Perseid, Eta Aquarid, and Southern Taurid meteor showers are hampered by moonlight, but the moon is largely out of the way for the Quadrantid (January 3) and Geminid showers (December 13), normally the year's best.

There are four eclipses in 1993, two of the sun and two of the moon. The total lunar eclipse of June 4 favors Hawaii, but for the westernmost part of North America totality begins as the moon sets. The lunar eclipse of November 29, also total, should be excellent for all observers in the western hemisphere.

1993 Sky Almanac

January 1 First quarter moon, 3:39 UT.
 3 Quadrantid meteor shower peaks this morning (14h UT).
 4 Earth closest to sun (perihelion), 3h UT. Distance: 91,406,000 miles or 147,096,900 kilometers.
 7 Mars (-1.5) at opposition. It rises in the northeast around sunset and is visible all night long. See page 54. Moon near Mars this evening.
 8 Full moon, 12:38 UT.
 14 Moon near Jupiter this morning.
 15 Last quarter moon, 4:02 UT.

19 Venus (-4.4) reaches greatest eastern elongation (47°). This is one of its best apparitions, and it rises even higher until February 4. Look high in the southwest about forty-five minutes after sunset. See page 18.
22 New moon, 18:28 UT.
23 Mercury at superior conjunction (not visible).
26 Moon near Venus this evening.
30 First quarter moon, 23:20 UT. Mars (-0.9) attains its greatest northern declination (+27°). The planet passes directly overhead for evening viewers located at that northern latitude (for example, central Florida and southern Texas).

February 3 Moon near Mars this evening.
 4 Venus (-4.5) at its best. Look high in the southwest forty-five minutes after sunset. Venus now sets nearly four hours after the sun and remains easily visible until mid-March.
 6 Full moon, 23:56 UT.
 9 Saturn in conjunction with the sun (not visible).
 10 Moon near Jupiter this morning.
 13 Last quarter moon, 14:58 UT.
 17 Mercury (-0.9) emerges from the sun's glare and into evening twilight. Begin looking for it low in the west about thirty minutes after sunset, below and to the right of brilliant Venus (-4.6). Mercury reaches its best altitude on the 21st.
 21 New moon, 13:06 UT. Mercury (-0.5) reaches greatest eastern elongation (18°). The planet is now at its best and remains well placed until the 25th. Watch for the young moon as it passes Mercury and Venus (-4.5) over the next few days. See page 21.

March 1 First quarter moon, 15:47 UT.
 3 Moon near Mars this evening.
 8 Full moon, 9:47 UT.
 9 Mercury at inferior conjunction (not visible).
 10 Moon near Jupiter this evening.
 15 Last quarter moon, 4:17 UT.
 20 Spring begins (vernal equinox, 14:42 UT). Moon near Saturn this morning.
 23 New moon, 7:16 UT.
 30 Jupiter (-2.5) at opposition. It rises in the east around sunset and is visible all night long. See page 58.
 31 First quarter moon, 4:11 UT. Moon near Mars this evening.

April 1 Venus at inferior conjunction (not visible).
 5 Mercury (+0.4) is at greatest western elongation (28°). Poor. Moon near Jupiter this evening.
 6 Full moon, 18:45 UT.
 13 Last quarter moon, 19:39 UT.
 16 Moon near Saturn this morning.
 19 Moon near Venus this morning. Venus (-4.4) emerges from the sun's glare and begins its rise into morning twilight. It reaches its best altitude August 4. Start looking for it low in the east in the hour before dawn.

21 New moon, 23:49 UT.
22 Lyrid meteor shower peaks this morning.
28 Moon near Mars this evening.
29 First quarter moon, 12:42 UT.

May 2 Moon near Jupiter tonight and tomorrow evening.
4 Eta Aquarid meteor shower peaks tonight. Moon interferes.
6 Full moon, 3:35 UT.
13 Last quarter moon, 12:20 UT.
14 Moon near Saturn this morning.
16 Mercury at superior conjunction (not visible).
17 Moon near Venus this morning.
21 New moon, 14:08 UT.
Partial solar eclipse begins 12:19 UT, visible from western U.S. and most of Canada.
26 Moon near Mars this evening.
28 First quarter moon, 18:22 UT.
30 Moon near Jupiter this evening.
31 Mercury (-0.8) emerges from the sun's glare and into evening twilight. Begin looking for it low in the northwest about thirty minutes after sunset. This is one of Mercury's best appearances, but the only guide in the vicinity is fading Mars (+1.4).

June 4 Full moon, 13:03 UT.
Total lunar eclipse. Moon contacts umbra at 11:11 UT — see page 39 for details.
10 Moon near Saturn this morning.
Venus (-4.3) reaches greatest western elongation (46°). It gains altitude until August 4 and remains easy to see until late October.
12 Last quarter moon, 5:37 UT.
Mercury (+0.2) at its best, low in the northwest thirty minutes after sunset. Now 14° above the horizon, it remains well placed until the 23rd. Fading Mars (+1.4) is the only nearby guide, above and to the left of Mercury.
16 Moon near Venus this morning.
17 Mercury (+0.5) reaches greatest eastern elongation (25°).
20 New moon, 1:54 UT.
21 Summer begins (summer solstice, 9:01 UT). This is the longest day of the year.
Moon near Mercury low in the west this evening.
23 Moon near Mars tonight and tomorrow evening.
26 First quarter moon, 22:45 UT.
Moon near Jupiter this evening.

July 3 Full moon, 23:46 UT.
4 Earth farthest from sun (aphelion), 22h UT. Distance: 94,509,500 miles or 152,091,200 kilometers.
7 Moon near Saturn this morning.
11 Last quarter moon, 22:50 UT.
15 Mercury at inferior conjunction (not visible).
16 Moon near Venus this morning.
19 New moon, 11:25 UT.
22 Moon near Mars this evening.
23 Moon near Jupiter this evening.
26 First quarter moon, 3:26 UT.

28 Southern Delta Aquarid meteor shower peaks this morning.

August 2 Full moon, 12:11 UT.
3 Moon near Saturn this morning.
4 Venus (-4.0) at its best, rising in the northeast three hours before dawn. Although this is not one of its best apparitions, Venus remains easily visible until late October.
Mercury (+0.2) is at greatest western elongation (19°). A poor predawn showing — but if you're out looking at Venus thirty minutes before sunrise during the next few days, try spotting Mercury hugging the horizon below and to its left.
10 Last quarter moon, 15:20 UT.
12 Perseid meteor shower peaks this morning. Moon interferes.
15 Moon near Venus this morning.
17 New moon, 19:29 UT.
19 Saturn (+0.3) at opposition. It rises in the southeast around sunset and is visible all night long.
20 Moon near Mars and Jupiter this evening.
24 First quarter moon, 9:59 UT.
29 Mercury at superior conjunction (not visible).
30 Moon near Saturn this evening.

September 1 Full moon, 2:34 UT.
9 Last quarter moon, 6:27 UT.
14 Moon below Venus this morning.
16 New moon, 3:12 UT.
22 First quarter moon, 19:33 UT.
23 Fall begins (autumnal equinox, 0:24 UT).
27 Moon near Saturn this evening.
30 Full moon, 18:54 UT.

October 8 Last quarter moon, 19:37 UT.
13 Moon near Venus this morning.
14 Mercury (+0.0) reaches greatest eastern elongation (25°). Poor.
15 New moon, 11:37 UT.
18 Jupiter in conjunction with the sun (not visible).
21 Orionid meteor shower peaks tonight.
22 First quarter moon, 8:53 UT.
23 Moon near Saturn this evening.
30 Full moon, 12:38 UT.

November 2 Southern Taurid meteor shower peaks tonight. Moon interferes.
6 Mercury at inferior conjunction (not visible). Mercury transits the sun for Asian observers. This event requires a telescope to observe as well as a safe method for viewing the sun. Mercury spends over ninety minutes silhouetted against the sun's disk beginning at 3:08 UT. Mercury's path across the sun is shown on page 26.
7 Last quarter moon, 6:37 UT.
8 Venus (-3.9) and Jupiter (-1.7) are closest this morning. Look low in the east thirty minutes before sunrise.
12 Moon near Jupiter this morning.
13 New moon, 21:35 UT.

13 Partial solar eclipse begins 19:46 UT, visible from New Zealand, Southeastern Australia, and Southern Chile.

16 Mercury (+0.3) emerges from the sun's glare and into morning twilight. Begin looking for it in the southeast below and to the left of Jupiter (-1.7) thirty minutes before dawn. See page 21.

17 Leonid meteor shower peaks tonight.

20 Moon near Saturn this evening.

21 First quarter moon, 2:04 UT.

22 Mercury (-0.5) is at greatest western elongation (19°). The planet is at its best and remains well placed for observing until month's end. Look along the southeastern horizon thirty minutes before dawn. See page 20.

29 Full moon, 6:32 UT.
Total lunar eclipse. Moon contacts umbra at 4:40 UT — see page 40 for details.

December 6 Last quarter moon, 15:50 UT.

10 Moon near Jupiter this morning.

13 New moon, 9:28 UT.
Geminid meteor shower peaks tonight (5h UT on the 14th).

17 Moon near Saturn this evening.

20 First quarter moon, 22:26 UT.

21 Winter begins (winter solstice, 20:27 UT). This is the shortest day of the year.

27 Mars in conjunction with the sun (not visible).

28 Full moon, 23:07 UT.

1994 Sky Summary

The year begins quietly, with only Saturn appearing in the west as night falls. Before it disappears into evening twilight, Saturn is met by **Mercury**, which rises to greet the Ringed Planet as February opens. This is the first of three good appearances by Mercury, but the planet's evening apparition in late May is outstanding — one of the best of the decade, in fact.

Spring brings the return of **Venus** to the evening sky, where it makes a rather poor showing. Though Venus is our evening star until late October, it may not catch your eye until the beginning of summer. During its twilight reign in the west, Venus receives embassies from Mercury in late May and Jupiter in early September.

Mars has one of its off years in 1994. The Red Planet spends the year drifting slowly through the stars of the morning sky. The only highlight is its appearance with Saturn shortly before sunrise in mid-March.

The slow-moving outer planets **Jupiter** and **Saturn** largely replay their 1993 appearances. Saturn slips into the morning sky in March and remains there until it reaches opposition in early September. Jupiter again begins the year in the south at sunrise, breaking into the evening sky with its opposition at the end of April. The "king of the planets" has only Venus for competition through the summer months.

This isn't a great year for meteors. Among the major showers, only the Perseids (August 12) and Geminids (December 14) are unobstructed by moonlight.

The highlight of the year will be a daytime event: the annular eclipse of the Sun on May 10. A large swath of the United States will enjoy the full extent of this fascinating phenomenon. Remember, this is a potentially hazardous event as well — please review the safe observing tips on page 31 before viewing the sun. A partial lunar eclipse follows this event at the end of May, and most of North America is favored. A total solar eclipse occurs in early November and favors central South America.

1994 Sky Almanac

January	2	Earth closest to the sun (perihelion), 6h UT. Distance: 91,407,800 miles or 147,099,700 kilometers.
	3	Mercury at superior conjunction (not visible). Quadrantid meteor shower peaks tonight (20h UT). Moon interferes.
	5	Last quarter moon, 00:01 UT.
	6	Moon near Jupiter this morning.
	11	New moon, 23:11 UT.
	14	Moon near Saturn this evening.
	17	Venus at superior conjunction (not visible).
	19	First quarter moon, 20:27 UT.
	27	Full moon, 13:24 UT.
February	1	Mercury (-0.8) emerges from the sun's glare and into evening twilight and remains well placed for observing until the 8th. Begin looking for it low in the southwest thirty minutes after sunset.
	2	Mercury (-0.8) close to Saturn (+0.9) tonight.
	3	Last quarter moon, 8:07 UT. Moon near Jupiter this morning.
	4	Mercury (-0.6) reaches greatest eastern elongation (18°).
	5	Mercury (-0.6) at its best, low in the southwest thirty minutes after sunset.
	10	New moon, 14:31 UT.
	18	First quarter moon, 17:49 UT.
	20	Mercury at inferior conjunction (not visible).
	21	Saturn in conjunction with the sun (not visible).
	26	Full moon, 01:16 UT.
March	2	Moon near Jupiter this morning.
	4	Last quarter moon, 16:55 UT.
	12	New moon, 7:06 UT.
	19	Mercury (+0.2) reaches greatest western elongation (28°). Poor.
	20	First quarter moon, 12:15 UT. Spring begins (vernal equinox, 20:29 UT).
	27	Full moon, 11:11 UT.
April	1	Venus (-3.9) emerges from the sun's glare and into evening twilight. Begin looking for it low in the northwest in the hour after sunset. Reaching its best altitude on June 7 — just 19° — forty-five minutes after sunset, this will be the planet's worst evening apparition in the period covered by this book.
	3	Last quarter moon, 2:55 UT.
	7	Moon near Saturn this morning.
	11	New moon, 00:18 UT.
	12	Moon near Venus this evening.
	19	First quarter moon, 2:35 UT.
	22	Lyrid meteor shower peaks this morning. Moon interferes.
	25	Full moon, 19:46 UT.
	26	Moon near Jupiter tonight.
	30	Mercury at superior conjunction (not visible). Jupiter (-2.5) at opposition. It rises in the southeast around sunset and is visible all night long.
May	2	Last quarter moon, 14:33 UT.
	4	Moon near Saturn this morning.
	5	Eta Aquarid meteor shower peaks this morning. Moon interferes.
	8	Moon near Mars this morning.
	10	New moon, 17:08 UT. Annular solar eclipse begins 14:12 UT — see map on page 33.
	12	Moon below Venus this evening.
	18	First quarter moon, 12:51 UT.
	23	Moon near Jupiter this evening.
	25	Full moon, 3:40 UT. Partial lunar eclipse. Moon contacts umbra 2:38 UT — see map on page 41.

27 Mercury (+0.2) at its best. Look for it below and to the right of brilliant Venus (-4.0), low in the northwest thirty minutes after sunset. Mercury remains well placed for viewing until June 8. See page 22.

30 Mercury (+0.5) at greatest eastern elongation (23°).

June 1 Last quarter moon, 4:03 UT. Moon near Saturn this morning.

6 Moon near Mars this morning.

7 Venus (-4.0) at its best, low in the west in the hour after sunset. Although it remains available for viewing until late August, this is its worst evening apparition in the period covered by this book. It now sets two-and-a-half hours after the sun. See page 19.

9 New moon, 8:28 UT.

11 Moon below Venus this evening.

16 First quarter moon, 19:57 UT.

19 Moon near Jupiter this evening.

21 Summer begins (summer solstice, 14:49 UT). This is the longest day of the year.

23 Full moon, 11:34 UT.

25 Mercury at inferior conjunction (not visible).

28 Moon near Saturn this morning

30 Last quarter moon, 19:31 UT.

July 5 Moon near Mars this morning. Earth farthest from sun (aphelion), 19h UT. Distance: 94,514,800 miles or 152,099,700 kilometers.

8 New moon, 21:39 UT.

11 Moon below Venus tonight.

16 First quarter moon, 1:13 UT. Moon near Jupiter this evening.

17 Mercury (+0.4) at greatest western elongation (21°). Poor.

22 Full moon, 20:17 UT.

25 Moon near Saturn this morning.

28 Southern Delta Aquarid meteor shower peaks this morning. Moon interferes.

30 Last quarter moon, 12:41 UT.

August 3 Moon near Mars this morning.

7 New moon, 8:46 UT.

10 Moon near Venus tonight.

12 Perseid meteor shower peaks tonight.

13 Moon near Jupiter this evening. Mercury at superior conjunction (not visible).

14 First quarter moon, 5:58 UT.

21 Full moon, 6:48 UT.

22 Moon near Saturn this morning.

24 Venus (-4.3) at greatest eastern elongation (46°), but it's preparing for a rapid exit. Before month's end it will appear less than 10° above the horizon forty-five minutes after sunset.

29 Last quarter moon, 6:41 UT.

September 1 Saturn (+0.5) at opposition. It rises around sunset and is visible all night long. Moon near Mars this morning.

5 New moon, 18:34 UT.

9 Moon near Jupiter this evening; Venus is nearby.

12 First quarter moon, 11:35 UT.

17 Moon near Saturn this evening.

19 Full moon, 20:02 UT.

23 Fall begins (autumnal equinox, 6:20 UT).

26 Mercury (+0.1) at greatest eastern elongation (26°). Poor.

28 Last quarter moon, 00:24 UT.

29 Moon near Mars this morning.

October 5 New moon, 3:56 UT.

7 Moon near Jupiter this evening.

11 First quarter moon, 19:18 UT.

14 Moon near Saturn tonight and tomorrow evening.

19 Full moon, 12:19 UT.

21 Mercury at inferior conjunction (not visible). Orionid meteor shower peaks tonight. Moon interferes.

27 Last quarter moon, 16:45 UT.

28 Moon near Mars this morning.

31 Mercury (+0.3) emerges into the morning sky, reaching its best altitude on November 6. Begin looking for it low in the southeast thirty minutes before sunrise.

November 2 Venus at inferior conjunction (not visible).

3 New moon, 13:37 UT. Total solar eclipse begins 11:05 UT, visible from central South America. Southern Taurid meteor shower peaks this morning.

6 Mercury (-0.5) at greatest western elongation (19°) and at its best low in the southeast thirty minutes before dawn.

10 First quarter moon, 6:15 UT.

11 Moon near Saturn this evening.

13 Venus (-4.3) emerges from the sun's glare to join Mercury (-0.7) in the morning twilight. Look for them both low in the southeast thirty minutes before dawn. See page 22.

17 Jupiter in conjunction with the sun (not visible). Leonid meteor shower peaks tonight. Moon interferes.

18 Full moon, 6:58 UT.

25 Moon near Mars this morning.

26 Last quarter moon, 7:05 UT.

30 Moon near Venus this morning.

December 2 New moon, 23:55 UT.

8 Moon near Saturn this evening.

9 First quarter moon, 21:08 UT.

14 Mercury at superior conjunction (not visible). Geminid meteor shower peaks this morning (12h UT).

18 Full moon, 2:18 UT.

22 Winter begins (winter solstice, 2:24 UT). This is the shortest day of the year. Venus (-4.6) at its best, rising nearly four hours before the sun. Bright Jupiter (-1.7) is below and to the left of Venus. Look for them high in the southeast in the hour before dawn.

23 Moon near Mars this morning.

25 Last quarter moon, 19:08 UT.

29 Moon between Venus and Jupiter this morning.

1995 Sky Summary

The evening sky is a bit weak in the planet department as the year begins; the morning sky is where the bright lights shine. The situation reverses by year's end.

Mercury starts the year in fine fashion, putting in one of its best appearances in mid-January. The best appearance of the year comes in April, when the planet returns to the evening sky. Of the three morning apparitions, only the July and October appearances offer much chance of seeing this elusive planet.

After a rather poor evening showing in 1994, **Venus** starts 1995 with a lackluster conclusion to the morning apparition it started the previous year. It won't enter the morning sky until late August, but you'll have a tough time spotting it after mid-April. Its return to the evening sky is nothing to brag about, either: it probably won't catch your eye until December.

Mars reaches an oppositon in mid-February — but it's the worst of the period covered in this book. However, it will be bright enough to outshine the stars of Leo, which it passes through this spring. After that, it goes through its usual lingering fade-out in the west.

The gap between **Jupiter** and **Saturn** has narrowed somewhat, but the year begins with Jupiter rising into the morning sky in Scorpius and Saturn loitering in the evening in the fall constellations. Jupiter, of course, is a summer planet this year, as it passes above Antares. For part of the year, Saturn is not the Ringed Planet, as the rings are presented edge-on to us and are invisible. This makes Saturn's September opposition an unusually faint one.

Of the year's meteor showers, two of the best, the Perseids and Geminids, are hampered by moonlight. The Quadrantids and Orionids are in fine shape, however, along with the lesser Eta and Delta Aquarids and the Leonids.

There are three eclipses this year. An annular solar eclipse in April is visible from the southern half of the Western Hemisphere, while the total solar eclipse of October is visible in the northern half of the Eastern Hemisphere. A partial lunar eclipse favoring western North America occurs in April.

1995 Sky Almanac

January 1   New moon, 10:57 UT.
       3   Quadrantid meteor shower peaks tonight (3h UT on the 4th).
       4   Earth closest to sun (perihelion), 11h UT. Distance: 91,407,900 miles or 147,099,900 kilometers.
       5   Moon near Saturn this evening.
       8   First quarter moon, 15:47 UT.
      13   Venus (-4.4) reaches greatest western elongation (47°). Although Venus has already passed its prime for this apparition, it remains an easy sight until mid-March.

14   Venus (-4.4) lies closest to Jupiter (-1.8) this morning. Look for the pair in the southeast forty-five minutes before sunrise. The waning moon joins the pair next week. See page 61.
16   Full moon, 20:27 UT.
19   Mercury (-0.6) reaches greatest eastern elongation (19°). It just manages to put in a decent appearance, staying above 10° only until the 22nd. Look for it low in the south-west about thirty minutes after sunset. Moon near Mars this morning.
24   Last quarter moon, 4:59 UT. Moon near Jupiter this morning.
27   Moon very close to Venus this morning. See page 61.
30   New moon, 22:48 UT.

February 1   Moon near Saturn this evening.
      3   Mercury at inferior conjunction (not visible).
      7   First quarter moon, 12:55 UT.
      12   Mars (-1.2) at opposition. It rises in the northeast at sunset and is visible all night long. See page 54.
      14   Moon near Mars this evening.
      15   Full moon, 12:16 UT.
      22   Last quarter moon, 13:05 UT.
      23   Moon near Jupiter this morning.
      26   Moon near Venus this morning.

March 1   Mercury (+0.1) reaches greatest western elongation (27°). Poor. New moon, 11:49 UT.
      6   Saturn in conjunction with the sun (not visible).
      9   First quarter moon, 10:15 UT.
      13   Moon near Mars this evening.
      17   Full moon, 1:26 UT.
      21   Spring begins (vernal equinox, 2:16 UT).
      22   Moon near Jupiter this morning.
      23   Last quarter moon, 20:11 UT.
      28   Moon near Venus this morning.
      31   New moon, 2:09 UT.

April 8   First quarter moon, 5:36 UT.
      10   Moon near Mars this evening.
      14   Mercury at superior conjunction (not visible).
      15   Full moon, 12:09 UT. Partial lunar eclipse. Moon contacts umbra at 11:41 UT, visible from Pacific Ocean, Australia, and eastern Asia.
      18   Moon near Jupiter this morning.
      22   Lyrid meteor shower peaks tonight. Moon interferes. Last quarter moon, 3:19 UT.
      25   Moon near Saturn this morning.
      27   Moon near Venus this morning.
      29   New moon, 17:37 UT. Annular solar eclipse begins 14:33 UT, visible from Ecuador, Peru, and Brazil.
      30   Begin looking for Mercury (-0.9) low in the northwest thirty minutes after sunset. It reaches its highest altitude on May 10 and remains well placed for viewing until the 21st.

May 5   Eta Aquarid meteor shower peaks tonight.
      7   First quarter moon, 21:45 UT.

8   Moon near Mars this evening.
10   Mercury (+0.2) at its best altitude. Look low in the northeast thirty minutes after sunset.
12   Mercury (+0.4) reaches greatest eastern elongation (22°).
14   Full moon, 20:49 UT.
16   Moon near Jupiter this morning.
21   Last quarter moon, 11:36 UT.
23   Moon near Saturn this morning.
27   Moon near Venus this morning.
29   New moon, 9:28 UT.

June 1   Jupiter at opposition (-2.6). It rises in the east at sunset and is visible all night long.
5   Mercury at inferior conjunction (not visible). Moon near Mars this evening.
6   First quarter moon, 10:27 UT.
11   Moon near Jupiter this evening.
13   Full moon, 4:04 UT.
19   Last quarter moon, 22:02 UT. Moon near Saturn this morning.
21   Summer begins (summer solstice, 20:35 UT).
28   New moon, 0:51 UT.
29   Mercury (+0.6) reaches greatest western elongation (22°). Poor.

July 3   Moon near Mars this evening.
4   Earth farthest from sun (aphelion), 2h UT. Distance: 94,516,500 miles or 152,102,400 kilometers.
5   First quarter moon, 20:04 UT.
9   Moon near Jupiter this evening.
12   Full moon, 10:50 UT.
17   Moon near Saturn this morning.
19   Last quarter moon, 11:11 UT.
27   New moon, 15:15 UT.
28   Mercury at superior conjunction (not visible).
29   Southern Delta Aquarid meteor shower peaks tonight.

August 1   Moon near Mars this evening.
4   First quarter moon, 3:17 UT.
5   Moon near Jupiter this evening.
10   Full moon, 18:17 UT.
12   Perseid meteor shower peaks tonight. Moon interferes.
13   Moon near Saturn this morning.
18   Last quarter moon, 3:04 UT.
20   Venus at superior conjunction (not visible).
26   New moon, 4:33 UT.
29   Moon near Mars this evening.

September 1   Moon near Jupiter this evening.
2   First quarter moon, 9:04 UT.
9   Mercury (+0.2) reaches greatest eastern elongation (27°). Poor. Full moon, 3:38 UT. Moon near Saturn this morning.
14   Saturn at opposition (+0.7). It rises in the east at sunset and is visible all night long.
16   Last quarter moon, 21:10 UT.
23   Fall begins (autumnal equinox, 12:15 UT).
24   New moon, 16:56 UT.
27   Moon near Mars this evening.
29   Moon near Jupiter this evening.

October 1   First quarter moon, 14:36 UT.
5   Mercury at inferior conjunction (not visible).
6   Moon near Saturn this evening.
8   Full moon, 15:53 UT.
16   Last quarter moon, 16:27 UT. Begin looking for Mercury (+0.2) low in the east thirty minutes before sunrise. By the 26th it's no longer easily visible.
20   Mercury (-0.5) reaches greatest western elongation (18°) and its best altitude for this apparition. Use the waning crescent moon as your guide to the planet. See page 22.
22   Orionid meteor shower peaks this morning.
24   New moon, 4:37 UT. Total solar eclipse begins 1:52 UT, visible from Iran, India, and Malaysia.
26   Moon near Mars and Jupiter this evening. See page 61.
30   First quarter moon, 21:18 UT.

November 2   Moon near Saturn this evening.
3   Southern Taurid meteor shower peaks tonight. Moon interferes.
7   Full moon, 7:21 UT.
15   Last quarter moon, 11:41 UT.
18   Leonid meteor shower peaks this morning.
22   New moon, 15:44 UT.
23   Mercury at superior conjunction (not visible). Moon near Jupiter and Venus this evening.
29   First quarter moon, 6:29 UT. Moon near Saturn this evening.

December 7   Full moon, 1:28 UT.
14   Geminid meteor shower peaks tonight. Moon interferes.
15   Last quarter moon, 5:33 UT. Venus (-4.0) emerges from the sun's glare and enters evening twilight. Begin looking for it low in the western sky in the hour after sunset. Remaining easily visible until late May, this is the planet's best evening apparition in the period covered by this book.
18   Jupiter in conjunction with the sun (not visible).
22   New moon, 2:24 UT. Winter begins (winter solstice, 8:19 UT). This is the shortest day of the year.
24   Moon near above Venus this evening.
27   Moon near Saturn this evening.
28   First quarter moon, 19:07 UT.

1996 Sky Summary

The beginning of the year belongs to **Venus**, the queen of the planets. After two lackluster years, Venus blazes high in the evening sky for the first half of the year. Venus returns to the morning sky in July, making a fine appearance there as well. Her suitors include Saturn in February and Mars (twice!) in the morning sky.

Mercury, always difficult, is a bit more so than usual this year. The planet starts the year fairly well, with an average appearance in the evening, and a better one in April. Mercury makes a half-hearted morning appearance in early October.

You might catch **Mars** in the evening sky in January, but 1996 is basically one of those years when the Red Planet "recovers" from its previous year's opposition by lingering in the morning sky.

Jupiter is once again a summer planet, appearing this year in Sagittarius. **Saturn** is not so far off now, lying beneath the Great Square of Pegasus, in the constellation Pisces.

Meteor-lovers can look forward to good shows by the Perseids and Geminids. The other strong shower, the Quadrantids, don't fare so well. Why not sneak a peek at the Leonids, since the moon's out of the way? Its parent comet is still a couple of years from its scheduled arrival — will the shower be better than usual this year?

We have two sets of eclipses this year. In April a total eclipse of the moon begins before moonrise in most parts of the United States. A partial solar eclipse follows two weeks later — but only penguins will enjoy it (and darned few of them!). The other eclipse set starts with September's total eclipse of the full moon that's conveniently timed for North Americans. When the moon is new two weeks later, there's a partial solar eclipse that favors Europe.

1996 Sky Almanac

January 2 Mercury (-0.6) reaches greatest eastern elongation (19°).
4 Earth closest to the sun (perihelion), 7h UT. Distance: 91,400,400 miles 147,087,900 kilometers.
Quadrantid meteor shower peaks this morning (9h UT). Moon interferes.
5 Full moon, 20:52 UT.
13 Last quarter moon, 20:47 UT.
18 Moon near Jupiter this morning. Mercury at inferior conjunction (not visible).
20 New moon, 12:51 UT.
22 Moon near Venus this evening.
23 Moon near Saturn this evening.
27 First quarter moon, 11:15 UT.

February 2 Venus (-4.1) closest to Saturn (+1.2) this evening. Look for the pair in the southwest in the hour after sunset.
4 Full moon, 15:59 UT.

11 Mercury (+0.0) reaches greatest western elongation (26°). Poor.
12 Last quarter moon, 8:38 UT.
15 Moon near Jupiter this morning.
18 New moon, 23:31 UT.
20 Moon near Saturn this evening.
21 Moon near Venus this evening.
26 First quarter moon, 5:53 UT.

March 4 Mars in conjunction with the sun (not visible).
5 Full moon, 9:24 UT.
12 Last quarter moon, 17:16 UT.
14 Moon near Jupiter this morning.
17 Saturn in conjunction with the sun (not visible).
19 New moon, 10:46 UT.
20 Spring begins (vernal equinox, 8:04 UT).
22 Moon near Venus this evening.
27 First quarter moon, 1:32 UT.
28 Mercury at superior conjunction (not visible).
31 Venus (-4.3) at its best, high in the west in the hour after sunset. It now sets nearly four hours after the sun and will remain an easy object until late May.

April 1 Venus (-4.3) reaches greatest eastern elongation (46°).
4 Full moon, 0:08 UT.
Total lunar eclipse. Moon contacts umbra at 22:21 UT — see page 42 for details.
10 Last quarter moon, 23:37 UT.
Moon near Jupiter this morning.
14 Mercury (-0.9) emerges from the sun's glare and enters evening twilight, remaining well placed for observation for the remainder of the month. Look for the planet in the western sky thirty minutes after sunset, below and to the right of brilliant Venus (-4.4). See page 23.
15 Moon near Saturn this morning.
17 New moon, 22:50 UT.
Partial solar eclipse begins 20:31 UT, visible from New Zealand and Antarctica.
19 Moon near Mercury this evening.
21 Lyrid meteor shower peaks tonight.
Moon near Venus this evening.
23 Mercury (+0.2) reaches greatest eastern elongation (20°).
25 First quarter moon, 20:42 UT.

May 3 Full moon, 11:49 UT.
4 Eta Aquarid meteor shower peaks tonight. Moon interferes.
5 Venus reaches greatest northern declination during the twentieth century (+28°).
7 Moon near Jupiter this morning.
10 Last quarter moon, 5:05 UT.
13 Moon near Saturn this morning.
15 Mercury at inferior conjunction (not visible).
17 New moon, 11:47 UT.
19 Moon near Venus this evening.
25 First quarter moon, 14:15 UT.

June 1 Full moon, 20:48 UT.
4 Moon near Jupiter this morning.
8 Last quarter moon, 11:06 UT.
9 Moon near Saturn this morning.

10 Mercury (+0.6) reaches greatest western elongation (24°). Poor.
Venus at inferior conjunction (not visible).

13 Moon near Mars this morning; Mercury lies below them.

16 New moon, 1:37 UT.

21 Summer begins (summer solstice, 2:25 UT). This is the longest day of the year.

24 First quarter moon, 5:25 UT.

30 Venus (-4.4) enters morning twilight. It lies near Mars (+1.5) this morning. Look low in the eastern sky thirty minutes before sunrise.

July 1 Full moon, 3:59 UT.
Moon near Jupiter this evening.

4 Jupiter at opposition (-2.7). It rises in the southeast at sunset and remains visible all night long.

5 Earth farthest from sun (aphelion), 19h UT. Distance: 94,514,200 miles or 152,098,800 kilometers.
Venus (-4.4) is now an easy sight in the pre-dawn sky. It increases altitude until September 6 and remains a prominent morning object until mid-December. See page 18.

7 Last quarter moon, 18:56 UT.
Moon near Saturn this morning.

11 Mercury at superior conjunction (not visible).

12 Moon near Venus this morning. Look for Mars to their left.

15 New moon, 16:16 UT.

23 First quarter moon, 17:50 UT.

27 Southern Delta Aquarid meteor shower peaks tonight. Moon interferes.

28 Moon near Jupiter this evening.

30 Full moon, 10:37 UT.

August 3 Moon near Saturn this morning.

6 Last quarter moon, 5:26 UT.

10 Moon near Venus and Mars this morning.

12 Perseid meteor shower peaks this morning.

14 New moon, 7:35 UT.

20 Venus (-4.3) reaches greatest western elongation (46°).

21 Mercury (+0.3) reaches greatest eastern elongation (27°). Poor.

22 First quarter moon, 3:38 UT.

24 Moon near Jupiter this evening.

28 Full moon, 17:53 UT.

30 Moon near Saturn this morning.

September 4 Venus (-4.2) near Mars (+1.5) this morning — look in the east in the hour before dawn.
Last quarter moon, 19:07 UT.

8 Moon near Venus and Mars this morning.

12 New moon, 23:09 UT.

17 Mercury at inferior conjunction (not visible).

20 First quarter moon, 11:24 UT.
Moon near Jupiter this evening.

22 Fall begins (autumnal equinox, 18:01 UT).

26 Saturn (+0.5) at opposition. It rises in the east at sunset and is visible all night long.
Moon near Saturn tonight.

27 Full moon, 2:52 UT.
Total lunar eclipse. Moon enters umbra at 1:12 UT — see page 43 for details.

30 Mercury (+0.1) emerges from the sun's glare and enters the morning twilight. Begin looking low in the east about thirty minutes before sunrise.

October 3 Mercury (-0.4) reaches greatest western elongation (18°) and is at its morning best low in the east half an hour before dawn. Look higher up for Venus (-4.1) and Mars (+1.4). See page 23.

4 Last quarter moon, 12:06 UT.

7 Moon near Mars this morning.

8 Moon near Venus this morning.

12 New moon, 14:16 UT.
Partial solar eclipse begins 11:59 UT, visible from Europe and northern Africa.

18 Moon near Jupiter this evening.

19 First quarter moon, 18:10 UT.

21 Orionid meteor shower peaks tonight.

23 Moon near Saturn tonight.

26 Full moon, 14:12 UT.

November 1 Mercury at superior conjunction (not visible).

2 Southern Taurid meteor shower peaks tonight. Moon interferes.

3 Last quarter moon, 7:52 UT.

5 Moon near Mars this morning.

8 Moon near Venus this morning.

11 New moon, 4:17 UT.

14 Moon near Jupiter this evening.

17 Leonid meteor shower peaks this morning.

18 First quarter moon, 1:10 UT.

20 Moon near Saturn this evening.

25 Full moon, 4:11 UT.

December 3 Last quarter moon, 5:07 UT.
Moon near Mars this morning.

8 Moon near Venus this morning.

10 New moon, 16:58 UT.

12 Moon near Jupiter this evening.

13 Geminid meteor shower peaks tonight (1h UT on the 14th).

15 Mercury (-0.5) reaches greatest eastern elongation (21°). Poor.

17 First quarter moon, 9:31 UT.
Moon near Saturn this evening.

21 Winter begins (winter solstice, 14:07 UT). This is the shortest day of the year.

24 Full moon, 20:41 UT.

1997 Sky Summary

The sky seems quiet as the year begins — Saturn alone lies in the evening sky, dim Mars and sinking Venus in the predawn sky. Mercury sneaks into the morning twilight early in the year for a brief appearance near Venus.

It's not until spring that things begin to pick up. **Mars** comes to opposition in mid-March. This is only slightly better than the 1995 opposition, which was the worst of the decade, but the Red Planet will certainly be conspicuous among the stars of Virgo. In late March **Mercury** enters evening twilight for its best appearance of the year. Mercury's only other decent appearance comes in September mornings.

Venus seems to skip along the westerly horizon all spring and summer, never gaining much altitude. It's not until late fall that it attains enough altitude to be noticeable. It'll be bright enough to spark a few UFO reports, though!

Jupiter continues to close the gap between itself and **Saturn**. They both lie in autumn constellations now (Jupiter in Capricornus, Saturn in Pisces), and so will be morning objects most of the year.

In the meteor department: This year's best bet is the Perseid shower. The other strong showers, the December Geminids and the January Quadrantids, are washed out by moonlight. Those hoping for enhanced Leonid activity will likely be disappointed — the moon is full at the shower's peak.

This year's best eclipse of the moon occurs in March: a partial eclipse whose deep umbral phase favors all but extreme northwestern North America. A September total eclipse happens during daylight hours for North Americans. Solar eclipse are unexciting affairs this year. The total eclipse of March occurs mostly north of the Arctic Circle above Asia. A deep partial eclipse of the sun will be seen by Australians in September.

1997 Sky Almanac

January	1	Moon near Mars this morning. Earth closest to sun (perihelion), 23h UT. Distance: 91,404,600 miles or 147,094,600 kilometers.
	2	Mercury at inferior conjunction (not visible). Last quarter moon, 1:46 UT.
	3	Quadrantid meteor shower peaks this morning 10h UT. Moon interferes.
	7	Moon near Venus this morning.
	9	New moon, 4:26 UT.
	12	Rising Mercury (+0.5) lies near sinking Venus (-3.9) this morning. Look for the pair very low in the southeast thirty minutes before dawn. Otherwise this is a poor apparition for Mercury,which remains well placed only until the 20th. See page 23.
	13	Moon near Saturn this evening.
	15	First quarter moon, 20:03 UT.
	19	Jupiter in conjunction with the sun (not visible).
	23	Full moon, 15:12 UT.
	24	Mercury (-0.1) reaches greatest western elongation (24°). Poor.
	28	Moon near Mars this morning.
	31	Last quarter moon, 19:41 UT.
February	7	New moon, 15:08 UT.
	10	Moon near Saturn this evening.
	14	First quarter moon, 8:58 UT.
	22	Full moon, 10:28 UT.
	25	Moon near Mars this morning.
March	2	Last quarter moon, 9:39 UT.
	6	Moon near Jupiter this morning.
	9	New moon, 1:15 UT. Total solar eclipse begins at 23:17 UT, visible from Mongolia and Russia.
	11	Mercury at superior conjunction (not visible).
	16	First quarter moon, 0:07 UT.
	17	Mars (-1.3) at opposition. It rises in the east at sunset and is visible all night long. See page 55.
	20	Spring begins (vernal equinox, 13:56 UT).
	22	Moon near Mars tonight.
	24	Full moon, 4:46 UT. Partial lunar eclipse. Moon enters umbra at 2:58 UT — for details see page 44.
	29	Mercury (-0.9) emerges from the sun's glare and enters evening twilight. Begin looking for it low in the west thirty minutes after sunset.
	30	Saturn in conjunction with the sun (not visible).
	31	Last quarter moon, 19:39 UT.
April	2	Venus at superior conjunction (not visible).
	3	Moon near Jupiter this morning.
	5	Mercury (-0.1) at its best low in the west thirty minutes after sunset. It remains well placed through the 12th.
	6	Mercury (-0.1) reaches greatest eastern elongation (19°).
	7	New moon, 11:03 UT.
	14	First quarter moon, 17:01 UT.
	18	Moon near Mars this evening.
	22	Lyrid meteor shower peaks this morning. Moon interferes. Full moon, 20:35 UT.
	25	Mercury at inferior conjunction (not visible).
	30	Last quarter moon, 2:37 UT. Moon near Jupiter this morning.
May	4	Moon near Saturn this morning.
	5	Eta Aquarid meteor shower peaks this morning.
	6	New moon, 20:48 UT.
	14	First quarter moon, 10:56 UT.
	16	Moon near Mars this evening.
	22	Mercury reaches greatest western elongation (25°). Poor. Ful moon, 9:14 UT.
	28	Moon near Jupiter this morning.
	29	Last quarter moon, 7:52 UT.

June 1   Moon near Saturn this morning.
     5   New moon, 7:05 UT.
    13   First quarter moon, 4:53 UT.
        Moon near Mars this evening.
    20   Full moon, 19:10 UT.
    21   Summer begins (summer solstice, 8:21 UT).
    24   Moon near Jupiter this morning.
    25   Mercury at superior conjunction (not visible).
    27   Last quarter moon, 12:43 UT.
    28   Moon near Saturn this morning.

July 4   Earth farthest from sun (aphelion), 19h UT. Distance: 94,517,600 miles or 152,104,200 kilometers.
        New moon, 18:41 UT.
     6   Moon near Venus this evening.
    11   Moon near Mars this evening.
    12   First quarter moon, 21:45 UT.
    20   Full moon, 3:21 UT.
    21   Moon near Jupiter this morning.
    25   Moon near Saturn this morning.
    26   Last quarter moon, 18:30 UT.
    27   Although this is a poor apparition for Mercury (+0.1), it lies near Venus (-3.9) tonight. Catch the pair low in the west thirty minutes after sunset. Southern Delta Aquarid meteor shower peaks tonight. Moon interferes.

August 3   New moon, 8:15 UT.
     4   Mercury reaches greatest eastern elongtion (27°). Poor.
     5   Moon near Venus this evening.
     9   Jupiter (-2.8) at opposition. It rises in the east at sunset and is visible all night long. Moon near Mars this evening.
    11   First quarter moon, 12:43 UT.
    12   Perseid meteor shower peaks tonight.
    17   Moon near Jupiter tonight.
    18   Full moon, 10:57 UT.
    22   Moon near Saturn this morning.
    25   Last quarter moon, 2:24 UT.
    31   Mercury at inferior conjunction (not visible).

September 1   New moon, 23:53 UT.
     2   Partial solar eclipse begins at 21:44 UT, visible from Australia and New Zealand.
     5   Moon near Venus this evening.
     6   Moon to the right of Mars this evening.
     7   Moon above Mars this evening.
    10   First quarter moon, 1:33 UT.
    13   Moon near Jupiter this evening.
    14   Mercury (+0.2) emerges from the sun's glare and enters morning twilight, remaining well placed for observation until the 20th. Look low in the east thirty minutes before dawn.
    16   Mercury (-0.2) reaches greatest western elongation (18°). Full moon, 18:52 UT. Total lunar eclipse. Moon enters umbra at 17:08 UT, visible from Asia, Africa, and Europe.
    17   Mercury (-0.3) at its best low in the east thirty minutes before sunrise.
    18   Moon near Saturn this morning.
    22   Fall begins (autumnal equinox, 23:57 UT).

September 23   Last quarter moon, 13:37 UT.

October 1   New moon, 16:53 UT.
     5   Moon above Venus and Mars this evening.
     9   First quarter moon, 12:23 UT.
    10   Saturn (+0.2) at opposition. It rises in the east at sunset and is visible all night long. Moon near Jupiter this evening.
    13   Mercury at superior conjunction (not visible).
    15   Moon near Saturn tonight.
    16   Full moon, 3:47 UT.
    21   Orionid meteor shower peaks tonight. Moon interferes.
    23   Last quarter moon, 4:50 UT.
    26   Brilliant Venus (-4.3) lies near sinking Mars (+1.1) this evening. Look for them low in the southwest in the hour after sunset.
    31   New moon, 10:02 UT.

November 2   Southern Taurid meteor shower peaks tonight.
     3   Moon near Mars this evening.
     4   Moon near Venus this evening.
     6   Venus (-4.4) reaches greatest eastern elongation (47°). It reaches its highest altitude on December 9 — just 19° forty-five minutes after sunset — and remains available for easy observation through early January.
     7   First quarter moon, 21:44 UT. Moon near Jupiter this evening.
    11   Moon near Saturn this evening.
    14   Full moon, 14:13 UT.
    17   Leonid meteor shower peaks tonight. Moon interferes.
    21   Last quarter moon, 23:59 UT.
    28   Mercury (-0.4) reaches greatest eastern elongation (22°). Poor.
    30   New moon, 2:15 UT.

December 2   Moon near Mars this evening.
     3   Moon near Venus this evening.
     4   Moon near Jupiter this evening.
     7   First quarter moon, 6:11 UT.
     8   Moon near Saturn this evening.
     9   Venus (-4.4) at its best tonight, hovering low in the southwest and setting about three hours after the sun.
    14   Full moon, 2:38 UT. Geminid meteor shower peaks this morning (7h UT).
    17   Mercury at inferior conjunction (not visible).
    21   Venus (-4.6) and Mars (+1.2) are close this evening. Look low in the southwest thirty minutes after sunset. Bright Jupiter (-2.1) hovers nearby, and the moon joins the grouping just before year's end. See page 56. Winter begins (winter solstice, 20:09 UT). Last quarter moon, 21:44 UT.
    28   Mercury (+0.3) emerges from the sun's glare and enters morning twilight. Look for it low in the southeast thirty minutes before dawn. Moon below and to the left of Mercury this morning.
    29   New moon, 16:58 UT.
    31   Moon near Venus and Mars this evening.

1998 Sky Summary

The year begins with four planets in the west at sunset: Saturn, Jupiter, Venus, and Mars.

This is a poor year for **Venus**, which slips into the morning sky in mid-January, then peaks in the morning sky in February. The rest of the year it slips back toward the sun, in a slow drawn-out descent.

Mercury has its best appearances at the beginning and end of the year, in the morning sky. Ordinarily, we see two good apparitions and several very poor ones. This year, we have mediocre apparitions in March and July (evening), and September (morning).

Mars, of course, was at opposition last year, and so fades slowly in the west before starting one of its "slow burns" in the morning sky of late spring. "Wait'll next year," as sports fans are wont to say.

Jupiter and **Saturn**, as mentioned above, are evening stars in the early part of the year. Jupiter, in Capricornus, then slips into the morning sky, followed about a month later by Saturn. They reach opposition about a month apart in the fall.

Meteor fans get off to a good start with the Quandrantids, which should put on a good evening display on January 3. Our two best standing meteor showers, the Perseids and the Geminids, must compete with moonlight, unfortunately. The main event, however, just might be the return of the Leonid storm. The moon cooperates; will the meteors?

There are no good lunar eclipses in 1998, only a series of three penumbral eclipses. A total eclipse of the sun, visible in the northern tip of South America, will bring a partial eclipse to the southeast half of the continental United States in February. The total eclipse of August will be visible in Malaysia.

1998 Sky Almanac

January 1   Moon lies near bright Jupiter (-2.1) tonight. Begin looking southwest thirty minutes after sunset. Venus (-4.5) hovers closest to the horizon, and faint Mars (+1.2) can be found midway between the two bright planets. See page 56.

2   Mercury (-0.1) at its best, low in the southeast thirty minutes before dawn. It remains well placed until the 17th.

3   Quadrantid meteor shower peaks tonight (22h UT).

4   Earth closest to the sun (perihelion), 21h UT. Distance: 91,407,600 miles or 147,099,400 kilometers.

5   First quarter moon, 14:20 UT.
Moon near Saturn this evening.

6   Mercury (-0.2) reaches greatest western elongation (23°).

12   Full moon, 17:25 UT.

16   Venus at inferior conjunction (not visible).

20   Last quarter moon, 19:41 UT.

Jupiter (-2.0) very close to Mars (+1.2) tonight. Begin looking for them low in the southwest thirty minutes after sunset.

28   New moon, 6:02 UT.

29   Moon near Mars this evening.

February 1   Venus (-4.5) emerges from the sun's glare and enters morning twilight. This is a poor apparition. Venus reaches its best altitude on the 23rd — just 16° forty-five minutes before sunrise — but until May remains an easy sight in the southeast in the hour before dawn. Moon near Saturn this evening.

3   First quarter moon, 22:54 UT.

11   Full moon, 10:24 UT.

19   Last quarter moon, 15:28 UT.

22   Mercury at superior conjunction (not visible).

23   Jupiter in conjunction with the sun (not visible).
Venus (-4.6) at its highest and brightest for this apparition, rising almost two hours ahead of the sun. This morning it's beside the waning moon —look southeast forty-five minutes before dawn.

26   New moon, 17:27 UT.
Total solar eclipse begins 14:50 UT — for details see page 35.

March 1   Moon near Saturn this evening.

5   First quarter moon, 8:42 UT.

13   Full moon, 4:35 UT.

14   Mercury (-0.9) emerges from the sun's glare and enters evening twilight. It remains well placed through the 24th. Look low in the west thirty minutes after sunset.

19   Mercury (-0.3) at its best. Look for Saturn (+0.6) nearby, slightly above and to the left of Mercury, low in the west thirty minutes after sunset.

20   Mercury (-0.2) reaches greatest eastern elongation (19°).
Spring begins (vernal equinox, 19:56 UT).

21   Last quarter moon, 7:38 UT.

24   Moon near Venus this morning.

27   Venus (-4.4) reaches greatest western elongation (46°).

28   New moon, 3:15 UT.

April 3   First quarter moon, 20:20 UT.

6   Mercury at inferior conjunction (not visible).

11   Full moon, 22:24 UT.

13   Saturn in conjunction with the sun (not visible).

19   Last quarter moon, 19:54 UT.

22   Lyrid meteor shower peaks tonight. Moon interferes.

23   Venus (-4.2) and Jupiter (-2.1) very close this morning. Look for the pair above the waning crescent moon, low in the east thirty minutes before dawn. See page 62.

26   New moon, 11:43 UT.

May 3   First quarter moon, 10:05 UT.

4   Mercury (+0.5) reaches greatest western elongation (27°). Poor.

5   Eta Aquarid meteor shower peaks this morning.
11 Full moon, 14:30 UT.
12 Mars in conjunction with the sun (not visible).
19 Last quarter moon, 4:36 UT.
21 Moon near Jupiter this morning.
22 Moon near Venus this morning.
23 Moon near Saturn this morning.
25 New moon, 19:34 UT.
29 Venus (-4.0) very close to Saturn (+0.6) this morning. Look for them low in the east thirty minutes before dawn. See page 62.

June 2   First quarter moon, 1:46 UT.
10 Mercury at superior conjunction (not visible). Full moon, 4:19 UT.
17 Last quarter moon, 10:40 UT. Moon near Jupiter this morning.
19 Moon near Saturn this morning.
21 Summer begins (summer solstice, 14:03 UT). Moon near Venus this morning.
24 New moon, 3:52 UT.

July 1   First quarter moon, 18:43 UT.
3   Mercury (-0.2) emerges from the sun's glare for a brief visit in the evening twilight. It remains well placed through the 11th.
4   Earth farthest from sun (aphelion), 0h UT. Distance: 94,512,300 miles or 152, 095,700 kilometers.
9   Full moon, 16:02 UT.
14 Moon near Jupiter this morning.
16 Last quarter moon, 15:15 UT.
17 Mercury (+0.5) reaches greatest eastern elongation (27°). Moon near Saturn this morning.
21 Moon near Venus this morning.
23 New moon, 13:45 UT.
28 Southern Delta Aquarid meteor shower peaks this morning
31 First quarter moon, 12:06 UT.

August 8   Full moon, 2:11 UT.
11 Moon near Jupiter this morning.
12 Perseid meteor shower peaks tonight. Moon interferes.
13 Mercury at inferior conjunction (not visible). Moon near Saturn this morning.
14 Last quarter moon, 19:49 UT.
19 Moon near Mars this morning.
22 New moon, 2:04 UT. Annular solar eclipse begins 23:10 UT, visible from Malaysia and Indonesia.
30 First quarter moon, 5:08 UT. Mercury (+0.3) emerges from the sun's glare and enters morning twilight. Look first for bright Venus (-3.9), low in the east thirty minutes before sunrise; Mercury hovers above it.
31 Mercury (-0.1) reaches greatest western elongation (18°).

September 1   Mercury (-0.3) at its best while Venus (-3.9) sinks toward the sun. Look low in the east thirty minutes before dawn. Try spotting faint Mars (+1.7) higher up. See page 24.
6   Full moon, 11:23 UT. Moon near Jupiter tonight.

9   Moon near Saturn this morning.
13 Last quarter moon, 1:59 UT.
16 Jupiter (-2.9) at opposition. It rises in the east at sunset and is visible all night long.
17 Moon near Mars this morning.
20 New moon, 17:02 UT.
23 Fall begins (autumnal equinox, 5:39 UT).
25 Mercury at superior conjunction (not visible).
28 First quarter moon, 21:12 UT.

October 3   Moon near Jupiter this evening.
5   Full moon, 20:13 UT.
7   Moon near Saturn this morning.
12 Last quarter moon, 11:12 UT.
16 Moon near Mars this morning.
20 New moon, 10:10 UT.
22 Orionid meteor shower peaks this morning.
23 Saturn (+0.0) at opposition. It rises in the east at sunset and is visible all night long.
28 First quarter moon, 11:47 UT.
30 Venus at superior conjunction (not visible).
31 Moon near Jupiter this evening.

November 3   Southern Taurid meteor shower peaks this morning. Moon interferes. Moon near Saturn tonight.
4   Full moon, 5:20 UT.
11 Mercury reaches greatest eastern elongation (23°). Poor. Last quarter moon, 0:29 UT.
13 Moon near Mars this morning.
17 Leonid meteor shower peaks today (23h UT). Will this be the year the great Leonid meteor storm recurs? Although the peak occurs during daytime for North American viewers, go out this morning and tomorrow morning to see if Leonid activity is enhanced.
19 New moon, 4:28 UT.
27 First quarter moon, 0:24 UT. Moon near Jupiter this evening.
30 Moon near Saturn this evening.

December 1   Mercury at inferior conjunction (not visible).
3   Full moon, 15:21 UT.
10 Last quarter moon, 17:55 UT.
11 Mercury (+0.4) enters morning twilight and remains favorably placed for observation through the 26th. Look low in the southeast thirty minutes before sunrise.
12 Moon near Mars this morning.
14 Geminid meteor shower peaks this morning (14h UT). Moon interferes.
16 Mercury (-0.2) below the waning moon this morning, low in the east thirty minutes before sunrise. See page 24.
17 Mercury (-0.3) at its best altitude this morning. Moon below and to the left of Mercury this morning.
18 New moon, 22:44 UT.
20 Mercury (-0.4) reaches greatest western elongation (22°).
22 Winter begins (winter solstice, 1:58 UT).
24 Moon below and to the right of Jupiter this evening.
26 First quarter moon, 10:47 UT.
27 Moon near Saturn this evening.

1999 Sky Summary

The year begins with Venus, Jupiter, and Saturn in the evening sky, while the morning twilight holds the more difficult planets Mars and Mercury.

Mercury has an unusually good year, with decent evening apparitions in March and July, and good morning appearances in August and especially December.

Venus puts in a fine stint as an evening star during the first half of 1999. It makes a quick exit in July and reappears in the morning sky by late August.

Late April brings an opposition of **Mars**, the best of the decade (though not the best of this book!). This is a good year to compare Mars to its rival, Antares in Scorpius, which it passes in mid-September.

Jupiter and **Saturn** are now close enough to appear together almost constantly. They lie in the "twilight zone" between the Great Square of Pegasus and the Hyades of Taurus. The fall months bring a delightful double retrograde loop, as the two approach their oppositions at the end of October (Jupiter) and the beginning of November (Saturn).

This could be stupendous year for meteors. The Perseids and Geminids are both free of moonlight, but best of all is the chance that the erratic Leonids will produce an intense meteor storm. We hope you've been sharpening your meteor-watching skills!

The year has little to offer in the way of eclipses. An annular eclipse will be visible in Australia in mid-February. After a good partial lunar eclipse in July, some of which will be visible throughout North America, the year's only total solar eclipse tracks from England to India in August.

1999 Sky Almanac

January	2	Full moon, 2:51 UT.
	3	Earth closest to sun (perihelion), 13h UT. Distance: 91,405,800 miles or 147,096,600 kilometers.
	4	Quadrantid meteor shower peaks this morning (5h UT). Moon interferes.
	9	Last quarter moon, 14:23 UT. Moon near Mars this morning.
	17	New moon, 15:47 UT.
	18	Moon near Venus this evening.
	21	Moon near Jupiter this evening.
	24	First quarter moon, 19:17 UT. Moon near Saturn this evening.
	25	Venus (-3.9) emerges from the sun's glare and enters evening twilight. The planet is at its best in May, when it sets three-and-a-half hours after the sun. Look for it low in the west in the hour after sunset.
	31	Full moon, 16:08 UT.
February	4	Mercury at superior conjunction (not visible).
	7	Moon near Mars this morning.
	8	Last quarter moon, 11:59 UT.
	16	New moon, 6:40 UT. Annular solar eclipse begins 3:52 UT, visible from Australia.
	17	Moon near Venus this evening.
	18	Moon near Jupiter this evening.
	20	Moon near Saturn this evening.
	23	Venus (-4.0) very close to Jupiter (-2.1) tonight. Look for the brilliant pair low in the west in the hour after sunset. First quarter moon, 2:44 UT.
	26	Begin looking for Mercury low in the west thirty minutes after sunset. It remains well placed for viewing through the 7th.
March	2	Full moon, 7:00 UT.
	3	Mercury (-0.4) reaches greatest eastern elongation (19°) and is at its best, the lowest member in a string of bright planets — Saturn (+0.5) is highest, then Venus (-4.0), Jupiter (-2.1), and Mercury. Look low in the west thirty minutes after sunset. See page 24.
	5	Mercury (+0.1) closest to Jupiter (-2.1) tonight. Look low in the west thirty minutes after sunset.
	7	Moon near Mars this morning.
	10	Last quarter moon, 8:42 UT.
	17	New moon, 18:49 UT.
	19	Mercury at inferior conjunction (not visible). Moon near Venus this evening.
	20	Venus (-4.0) closest to Saturn (+0.5) tonight. Look for the pair below and to the right of the young moon, low in the west in the hour after sunset.
	21	Spring begins (vernal equinox, 1:47 UT).
	24	First quarter moon, 10:19 UT.
	31	Full moon, 22:50 UT.
April	1	Jupiter in conjunction with the sun (not visible).
	3	Moon near Mars this morning.
	9	Last quarter moon, 2:52 UT.
	16	Mercury (+0.4) reaches greatest western elongation (28°). Poor. New moon, 4:23 UT.
	18	Moon near Venus this evening.
	22	Lyrid meteor shower peaks tonight. Moon interferes. First quarter moon, 19:03 UT.
	24	Mars (-1.7) at opposition. It rises in the east after sunset and is visible all night long.
	27	Saturn in conjunction with the sun (not visible).
	29	Moon near Mars tonight.
	30	Full moon, 14:56 UT.
May	5	Eta Aquarid meteor shower peaks tonight. Moon interferes.
	8	Last quarter moon, 17:29 UT. Moon near Jupiter this morning.
	11	Venus (-4.1) at its best for this apparition, hovering 30° above the western horizon forty-five minutes after sunset. It now sets three and a half hours after the sun. Venus remains an easy sight until mid-July.
	15	New moon, 12:06 UT.
	18	Moon near Venus this evening.
	22	First quarter moon, 5:35 UT.
	25	Mercury at superior conjunction (not visible).
	26	Moon near Mars this evening.
	30	Full moon, 6:41 UT.
June	7	Last quarter moon, 4:21 UT.
	10	Moon near Jupiter this morning.

11 Moon near Saturn this morning.
Venus (-4.3) reaches greatest eastern elongation (45°). This is its least greatest elongation in
the twentieth century.
Mercury (-0.6) emerges from the sun's glare
and enters evening twilight. Begin looking for
it below and to the right of brilliant Venus, low
in the west thirty minutes after sunset. It
remains well placed for the rest of the month.
13 New moon, 19:04 UT.
15 Moon near Mercury this evening.
16 Moon near Venus this evening.
20 First quarter moon, 18:14 UT.
21 Summer begins (summer solstice, 19:50 UT).
Mercury (+0.1) at its best in the evening sky,
low in the west thirty minutes after sunset.
22 Moon near Mars this evening.
28 Mercury (+0.5) reaches greatest eastern
elongation (26°).
Full moon, 21:38 UT.

July 6 Earth farthest from sun (aphelion), 23h UT.
Distance: 94,514,300 miles or 152,099,000
kilometers.
Last quarter moon, 11:58 UT.
7 Moon near Jupiter this morning.
8 Moon near Saturn this morning.
13 New moon, 2:25 UT.
15 Moon near Venus this evening.
20 First quarter moon, 9:02 UT.
Moon near Mars this evening.
26 Mercury at inferior conjunction (not visible).
28 Full moon, 11:26 UT.
Partial lunar eclipse. Moon enters umbra at
10:22 UT — for details see page 45.
Southern Delta Aquarid meteor shower peaks
tonight. Moon interferes.

August 4 Last quarter moon, 17:28 UT.
Moon near Jupiter this morning.
5 Moon near Saturn this morning.
11 New moon, 11:09 UT.
Total solar eclipse begins 8:26 UT — for
details see page 36.
12 Mercury (+0.5) emerges into morning twilight,
but remains favorably placed only until the
19th. Look low in the east thirty minutes
before dawn.
Perseid meteor shower peaks tonight.
14 Mercury (+0.1) reaches greatest western
elongation (19°).
18 Moon near Mars this evening.
19 First quarter moon, 1:48 UT.
20 Venus at inferior conjunction (not visible).
26 Full moon, 23:49 UT.
31 Moon near Jupiter this morning.

September 1 Moon near Saturn this morning.
2 Last quarter moon, 22:18 UT.
7 Moon near Venus this morning.
8 Mercury at superior conjunction (not visible).
9 New moon, 22:03 UT.
16 Moon near Mars this evening.
17 First quarter moon, 20:07 UT.
23 Fall begins (autumnal equinox, 11:33 UT).
25 Full moon, 10:52 UT.
27 Moon near Jupiter this morning.
28 Moon near Saturn this morning.

29 Venus (-4.6) emerges from the sun's glare and
into morning twilight. It reaches its best
altitude next month, but remains a prominent
morning object until late January 2000. Look
for it low in the east forty-five minutes before
dawn.

October 2 Last quarter moon, 4:03 UT.
5 Moon near Venus this morning.
9 New moon, 11:36 UT.
15 Moon near Mars this evening.
17 First quarter moon, 15:00 UT.
22 Orionid meteor shower peaks this morning.
Moon interferes.
23 Jupiter (-2.9) at opposition. It rises in the east
at sunset and remains visible all night long.
24 Mercury (-0.1) reaches greatest eastern
elongation (24°). Poor.
Full moon, 21:04 UT.
Moon near Jupiter tonight.
25 Moon near Saturn tonight.
29 Venus (-4.4) at its best in the morning sky,
high in the east in the hour before dawn.
30 Venus (-4.4) reaches greatest western elongation (46°).
31 Last quarter moon, 12:05 UT.

November 3 Southern Taurid meteor shower peaks tonight.
4 Moon near Venus this morning.
6 Saturn (-0.2) at opposition. It rises in the east
at sunset and remains visible all night long.
8 New moon, 3:54 UT.
13 Moon near Mars this evening.
15 Mercury at inferior conjunction (not visible).
Mercury transits the sun, but only Antarctica is
favored.
16 First quarter moon, 9:04 UT.
18 Leonid meteor shower peaks this morning
(5h UT). Will this be the year the great Leonid
meteor storm returns? Conditions are favorable
for viewers in eastern North America, with the
radiant low in the east and the moon low in
the west at the time of peak (midnight EST on
the **17th**).
20 Moon near Jupiter this evening.
21 Moon near Saturn this evening.
23 Full moon, 7:05 UT.
25 Mercury (+0.4) enters morning twilight. Look
for it low in the east about thirty minutes
before sunrise. It remains favorably placed
until December 10.
29 Last quarter moon, 23:19 UT.

December 3 Mercury (-0.4) reaches greatest western
elongation (20°).
4 The waning moon guides you to Venus (-4.2)
and Mercury (-0.5) over the next few days.
Look east in the hour before dawn. See page
24.
6 Moon below Mercury this morning.
7 New moon, 22:32 UT.
12 Moon near Mars this evening.
14 Geminid meteor shower peaks tonight (20h
UT).
16 First quarter moon, 0:51 UT.
17 Moon near Jupiter this evening.
18 Moon near Saturn this evening.
22 Full moon, 17:33 UT.
Winter begins (winter solstice, 7:45 UT).
29 Last quarter moon, 14:05 UT.

2000 Sky Summary

The year begins with the outer planets in the evening sky and the inner planets in the morning sky. Jupiter and Saturn are bright in the southeast at sunset, while Mars is somewhat fainter and to the south. Mercury lies in the morning sky, but isn't visible, while Venus is easy to spot in the southeast.

The year offers an excellent chance to spot **Mercury**, and a couple of lesser opportunities. The best appearance by far comes in June, when Mercury is in the evening sky. The best morning appearance comes in November. The February evening appearance and the July morning apparition offer decent chances to spot this elusive planet.

This isn't a banner year for **Venus**, which is only easily visible in the morning sky for about a month and a half, after which it lurks in the twilight glow until mid-June. When it enters the evening sky, it lies in twilight there until about the end of October.

Mars begins the year in one of its post-opposition "hangovers," hanging over the sunset horizon, growing dimmer by the week as our planet races away from it. The highlight of its year is its close approach to Jupiter and Saturn in April.

This is the year when **Jupiter** and **Saturn** have their mutual conjunction. As noted, the closest approach comes when the planets are too near the rising sun to see. However, when they do emerge from the morning twilight in mid-June, they'll still be a striking pair. They'll remain close all year, and approach one another at year's end, during their retrograde loops.

This is not a great year for the major meteor showers, with nearly full moons interfering with the Perseids and Geminids. However, there is a good possibility that the Quadrantids will be a fine morning shower.

The year starts with a fine, if chilly, total eclipse of the moon, whose timing is almost ideal — the entire umbral portion of the eclipse is visible throughout North America. That's about it for the year, however. The warmer July total lunar eclipse comes around sunrise, unfortunately. Two of the three partial solar eclipses are only visible from the extreme southern hemisphere; the third is worth watching for those in Alaska and northern Canada.

2000 Sky Almanac

January	3	Earth closest to sun (perihelion), 6h UT. Distance: 91,409,500 miles or 147,102,600 kilometers. Moon near Venus this morning.
	4	Quadrantid meteor shower peaks this morning (11h UT).
	6	New moon, 18:15 UT.
	10	Moon near Mars this evening.
	14	First quarter moon, 13:35 UT.
	15	Moon near Jupiter and Saturn this evening.
	16	Mercury at superior conjunction (not visible).
	21	Full moon, 4:42 UT.
		Total lunar eclipse. Moon contacts umbra at 3:01 UT — for details see page 46.
	28	Last quarter moon, 7:58 UT.
February	2	Moon near Venus this morning.
	5	New moon, 13:04 UT. Partial solar eclipse begins 10:56 UT, visible from Antarctica
	8	Moon near Mars this evening.
	10	Moon near Jupiter this evening.
	11	Mercury (-0.8) enters evening twilight. It's best on the 15th and remains favorably placed till the 18th. Look for it low in the west thirty minutes after sunset. Moon near Saturn this evening.
	12	First quarter moon, 23:22 UT.
	15	Mercury (-0.5) reaches greatest eastern elongation (18°) and is at its best, low in the west thirty minutes after sunset. Mars (+1.2) is not far off, above and to the left of Mercury.
	19	Full moon, 16:28 UT.
	27	Last quarter moon, 3:55 UT.
March	1	Mercury at inferior conjunction (not visible).
	6	New moon, 5:18 UT.
	8	Moon near Mars this evening.
	9	Moon near Jupiter this evening.
	10	Moon near Saturn this evening.
	13	First quarter moon, 7:00 UT.
	20	Full moon, 4:45 UT. Spring begins (vernal equinox, 7:37 UT).
	28	Mercury (+0.3) reaches greatest western elongation (28°). Poor. Last quarter moon, 0:22 UT.
April	4	New moon, 18:13 UT.
	6	Tonight, look right of the young moon for a triangle of planets formed by Saturn (+0.3), Jupiter (-2.0), and Mars (+1.4); Mars and Jupiter are closest tonight. Start looking low in the west thirty minutes after sunset. See page 56.
	11	First quarter moon, 13:31 UT.
	16	Mars (+1.5) closest to Saturn (+0.3) tonight. Look low in the west thirty minutes after sunset.
	18	Full moon, 17:43 UT.
	22	Lyrid meteor shower peaks this morning. Moon interferes.
	26	Last quarter moon, 19:31 UT.
May	4	New moon, 4:13 UT. Eta Aquarid meteor shower peaks tonight.
	8	Jupiter in conjunction with the sun (not visible).
	9	Mercury at superior conjunction (not visible).
	10	Saturn in conjunction with the sun (not visible). First quarter moon, 20:02 UT.
	18	Full moon, 7:36 UT.
	24	Mercury (-0.8) enters evening twilight for its best apparition of the year. At its best in early June, the planet remains favorably placed until June 16. Look low in the west thirty minutes after sunset.
	26	Last quarter moon, 11:56 UT.

31 Jupiter (-2.0) and Saturn (+0.2) are closest this morning (10h UT), but unfortunately the planets are very low in the east thirty minutes before sunrise. Watch them over the next few weeks as they become prominent morning objects. The pair will be a striking sight in November, when they reach opposition within nine days of one another.

June 2 New moon, 12:15 UT.
   4 Mercury (+0.2) reaches its best altitude tomorrow, but look for it below and to the right of the young moon this evening, low in the west thirty minutes after sunset. See page 25.
   9 Mercury (+0.5) reaches greatest eastern elongation (24°).
      First quarter moon, 3:30 UT.
   11 Venus at superior conjunction (not visible).
   16 Full moon, 22:28 UT.
   21 Summer begins (summer solstice, 1:49 UT).
   25 Last quarter moon, 1:01 UT.
   29 Moon below Saturn and Jupiter this morning.

July 1 New moon, 19:21 UT.
      Partial solar eclipse begins 18:07 UT, visible from southern Chile.
      Mars in conjunction with the sun (not visible).
   3 Earth farthest from the sun (aphelion), 24h UT. Distance: 94,516,500 miles or 152,102,400 kilometers.
   6 Mercury at inferior conjunction (not visible).
   8 First quarter moon, 12:54 UT.
   16 Full moon, 13:56 UT.
      Total lunar eclipse. Moon contacts umbra at 11:57 UT, visible from the Pacific Ocean, Australia, and eastern Asia..
   24 Last quarter moon, 11:03 UT.
   26 Moon near Saturn and Jupiter this morning.
   27 Mercury (+0.3) reaches greatest western elongation (20°).
      Southern Delta Aquarid meteor shower peaks tonight.
   29 Mercury (+0.0) just manages to make a brief appearance in morning twilight. Favorable only through the 31st; look for it near the waning moon this morning low in the east thirty minutes before sunrise. The bright planetary pair Jupiter (-2.2) and Saturn (+0.2) are now high in the east.
   31 New moon, 2:26 UT.
      Partial solar eclipse begins at 0:37 UT, visible from Siberia, Alaska, and northwestern Canada.

August 7 First quarter moon, 1:03 UT.
   12 Perseid meteor shower peaks this morning. Moon interferes.
   15 Full moon, 5:14 UT.
   22 Mercury at superior conjunction (not visible). Last quarter moon, 18:52 UT.
      Moon near Saturn and Jupiter this morning.
   29 New moon, 10:20 UT.

September 5 First quarter moon, 16:28 UT.
   13 Full moon, 19:38 UT.
   19 Moon near Saturn and Jupiter this morning.
   21 Last quarter moon, 1:29 UT.

22 Fall begins (autumnal equinox, 17:29 UT).
25 Moon near Mars this morning.
27 New moon, 19:54 UT.
29 Moon near Venus this evening.

October 5 First quarter moon, 11:00 UT.
   6 Mercury (+0.0) reaches greatest eastern elongation (26°). Poor.
   13 Full moon, 8:54 UT.
   16 Moon near Saturn this morning.
   17 Moon near Jupiter this morning.
   20 Last quarter moon, 8:00 UT.
   21 Orionid meteor shower peaks this morning. Moon interferes.
   24 Moon near Mars this morning.
   27 New moon, 7:59 UT.
   29 Moon near Venus this evening.
   30 Mercury at inferior conjunction (not visible).

November 2 Venus (-4.0) enters evening twilight for an excellent evening apparition — a replay of its evening display of 1992-93! See page 18. Highest on February 2, 2001, it remains an easy evening object until mid-March. Look low in the west in the hour after sunset.
      Southern Taurid meteor shower peaks tonight. Moon interferes.
   4 First quarter moon, 7:28 UT.
   9 Mercury (+0.3) enters morning twilight and remains favorably placed through the 22nd. Look low in the east thirty minutes before dawn.
   11 Full Moon, 21:16 UT.
      Moon near Saturn tonight.
   12 Moon near Jupiter tonight.
   14 Mercury (-0.4) at its best, low in the east thirty minutes before sunrise. Try to catch a glimpse of faint Mars (+1.7) higher up.
   15 Mercury (-0.5) reaches greatest western elongation (19°).
   17 Leonid meteor shower peaks tonight. Moon interferes.
   18 Last quarter moon, 15:26 UT.
   19 Saturn at (-0.4) opposition. It rises in the east at sunset and is visible all night long. See page 64.
   21 Moon near Mars this morning.
   25 New moon, 23:12 UT.
   28 Jupiter (-2.9) at opposition. It rises in the east at sunset and is visible all night long. Jupiter and Saturn (-0.3) gleam among the stars of Taurus all night long. See page 64.
   29 Moon near Venus this evening.

December 4 First quarter moon, 3:56 UT.
   9 Moon between Saturn and Jupiter this evening.
   11 Full moon, 9:04 UT.
   13 Geminid meteor shower peaks tonight. Moon interferes.
   18 Last quarter moon, 0:42 UT.
   20 Moon near Mars this morning.
   21 Winter begins (winter solstice, 13:39 UT).
   25 Mercury at superior conjunction (not visible). New moon, 17:23 UT.
      Partial solar eclipse begins at 15:27 UT — for details see page 37.
   29 Moon near Venus this evening.

2001 Sky Summary

The evening sky is again filled with planets as the new millennium begins. Venus, Jupiter and Saturn can all be seen in the evening on New Year's Day; Mercury joins them before January is over.

The best appearance for **Mercury** comes in May, in the evening sky. As usual, there's only one good morning appearance, this year in October, but it scarcely matches the May apparition.

Venus has a good year, ending an evening display as the year begins, it then heads quickly into the morning sky for a good appearance there. Toward the end of its morning apparition, it pairs up nicely with Mercury.

This year brings the best opposition of **Mars** in the period covered in this book. There will never be a fiercer competition between the planet and its stellar rival, Antares, the bright red heart of Scorpius, where the opposition takes place this summer.

As they conclude their retrograde loops early in the year, **Jupiter** and **Saturn** will make a fine pair in Taurus. Jupiter's greater orbital speed soon begins to tell, however, and the giant planet soon moves on to Gemini, leaving Saturn behind. The two won't meet again for another twenty years.

The Geminid shower figures to be the year's best, the erratic Leonids remaining a strong contender. The Perseids peak in moonlight, but the Orionids could put on a good show.

The first eclipse of the century, a lunar eclipse, occurs in daylight in North America — but Europe and Asia will have a good look. Southern Africans will see a total eclipse of the sun in June, which is followed by a partial lunar eclipse visible to Australians in early July. As the year ends, a total solar eclipse over the Pacific and Central America will give southern North Americans a chance to see a partial eclipse.

2001 Sky Almanac

January	2	First quarter moon, 22:33 UT.
	3	Quadrantid meteor shower peaks tonight (18h UT).
	4	Earth closest to the sun (perihelion), 9h UT. Distance: 91,406,300 miles or 147,097,300 kilometers.
	5	Moon near Saturn and Jupiter this evening.
	9	Full moon, 20:25 UT. Total lunar eclipse. Moon contacts umbra at 18:42 UT, visible from Asia, Africa and Europe.
	16	Last quarter moon, 12:36 UT.
	17	Venus (-4.4) reaches greatest eastern elongation (47°) and dominates the western sky in the hour after sunset. This is a replay of its 1992-93 apparition. See page 18. Moon near Mars this morning.
	24	New moon, 13:08 UT.

	25	Mercury (-0.8) enters evening twilight and remains well placed through February 1. Look for it low in the southwest thirty minutes after sunset. See page 25.
	28	Mercury (-0.6) reaches greatest eastern elongation (18°). Moon near Venus this evening.
	29	Mercury (-0.5) at its best low in the southwest thirty minutes after sunset.
February	1	First quarter moon, 14:03 UT. Moon near Saturn and Jupiter this evening.
	2	Venus (-4.5) at its best high in the evening sky and now sets nearly four hours after the sun. It remains favorably placed forty-five minutes after sunset until mid-March.
	8	Full moon, 7:13 UT.
	13	Mercury at inferior conjunction (not visible).
	15	Last quarter moon, 3:25 UT. Moon near Mars this morning.
	23	New moon, 8:22 UT.
	26	Moon near Venus this evening.
March	1	Moon near Saturn and Jupiter this evening.
	3	First quarter moon, 2:04 UT.
	9	Full moon, 17:24 UT.
	11	Mercury (+0.2) reaches greatest western elongation (27°). Poor.
	15	Moon near Mars this morning.
	16	Last quarter moon, 20:46 UT.
	20	Spring begins (vernal equinox, 13:32 UT).
	25	New moon, 1:22 UT.
	28	Moon near Saturn this evening.
	29	Moon near Jupiter this evening.
	30	Venus at inferior conjunction (not visible).
April	1	First quarter moon, 10:50 UT.
	8	Full moon, 3:23 UT.
	13	Moon near Mars this morning.
	15	Last quarter moon, 15:32 UT.
	21	Moon near Venus this morning.
	22	Lyrid meteor shower peaks this morning.
	23	Mercury at superior conjunction (not visible). New moon, 15:27 UT.
	25	Moon near Saturn this evening.
	26	Moon near Jupiter this evening.
	30	First quarter moon, 17:09 UT.
May	5	Eta Aquarid meteor shower peaks this morning. Moon interferes.
	7	Venus (-4.5) emerges into morning twilight. At its best in August, it remains a prominent morning object through late October. Look low in the east in the hour before dawn. Full moon, 13:54 UT.
	8	Mercury (-0.9) emerges into evening twilight and remains well placed through month's end. Look for bright Jupiter (-2.0) low in the west thirty minutes after sunset — Mercury lies below and to its right.
	10	Moon near Mars this morning.
	15	Last quarter moon, 10:12 UT.
	19	Moon near Venus this morning. Mercury (+0.2) at its best this evening, low in the west thirty minutes after sunset. Jupiter (-1.9) lies directly beneath it.

22 Mercury (+0.5) reaches greatest eastern elongation (22°).

23 New moon, 2:47 UT.

24 Moon near Mercury this evening.

25 Saturn in conjunction with the sun (not visible).

29 First quarter moon, 22:10 UT.

June 6 Full moon, 1:40 UT.
Moon near Mars tonight.

8 Venus (-4.3) reaches greatest western elongation (46°) and dominates the morning sky in the hour before dawn.

13 Mars (-2.3) at opposition. It rises in the east at sunset and is visible all night long. See page 56.

14 Jupiter in conjunction with the sun (not visible).
Last quarter moon, 3:29 UT.

16 Mercury at inferior conjunction (not visible).

17 Moon near Venus this morning.

19 Moon above Saturn this morning.

21 New moon, 11:59 UT.
Total solar eclipse begins 9:33 UT, visible from Angola, Zambia, Mozambique, and Madagascar.
Summer begins (summer solstice, 7:39 UT).

28 First quarter moon, 3:21 UT.

July 3 Moon near Mars this evening.

4 Earth farthest from the sun (aphelion), 14h UT. Distance: 94,507,300 miles or 152,087,600 kilometers.

5 Full moon, 15:05 UT.
Partial lunar eclipse. Moon contacts umbra at 13:35 UT, visible from Australia and eastern Asia.

9 Mercury (+0.5) reaches greatest western elongation (21°). Poor.

13 Last quarter moon, 18:46 UT.

15 Venus (-4.1) closest to Saturn (+0.2) this morning. Look for the pair in the east in the hour before dawn. See page 65.

17 Moon near Venus and Saturn this morning.

19 Moon near Jupiter this morning.

20 New moon, 19:45 UT.

27 First quarter moon, 10:09 UT.

28 Southern Delta Aquarid meteor shower peaks this morning. Moon interferes.

30 Moon near Mars this evening.

August 4 Full moon, 5:57 UT.

5 Mercury at superior conjunction (not visible). Venus (-4.0) closest to Jupiter (-2.0) this morning. Fainter Saturn (+0.1) hovers high above the pair. Look east in the hour before dawn. See page 65.

12 Perseid meteor shower peaks this morning. Moon interferes.
Last quarter moon, 7:54 UT.

14 Moon near Saturn this morning.

15 The waning moon lies within the morning chain of planets formed by Saturn (+0.1), Jupiter (-2.0), and Venus (-4.0). Look east half an hour before dawn.

16 Moon near Venus this morning.

19 New moon, 2:56 UT.

25 First quarter moon, 19:56 UT.

27 Moon near Mars this evening.

September 2 Full moon, 21:44 UT.

10 Last quarter moon, 19:01 UT.
Moon near Saturn this morning.

12 Moon near Jupiter this morning.

15 Moon near Venus this morning.

17 New moon, 10:28 UT.

18 Mercury (+0.1) reaches greatest eastern elongation (27°). Poor.

22 Fall begins (autumnal equinox, 23:06 UT).

24 First quarter moon, 9:32 UT.
Moon near Mars this evening.

October 2 Full moon, 13:50 UT.

7 Moon near Saturn this morning.

10 Last quarter moon, 4:21 UT.
Moon near Jupiter this morning.

14 Mercury at inferior conjunction (not visible).
Moon near Venus this morning.

16 New moon, 19:24 UT.

21 Orionid meteor shower peaks tonight.

23 Moon near Mars this evening.

24 First quarter moon, 2:59 UT.

25 Mercury (+0.1) enters morning twilight. Look low in the east thirty minutes before sunrise.

29 Mercury (-0.5) reaches greatest western elongation (19°) and is at its best. Mercury meets sinking Venus (-3.9) and the two remain less than 1° apart through November 7. Look low in the east thirty minutes before sunrise. See page 25.

November 1 Full moon, 5:42 UT.

3 Southern Taurid meteor shower peaks this morning. Moon interferes.
Moon near Saturn this morning.

6 Moon near Jupiter this morning.

8 Last quarter moon, 12:22 UT.

15 New moon, 6:41 UT.

17 Leonid meteor shower peaks tonight.

21 Moon near Mars this evening.

22 First quarter moon, 23:22 UT.

30 Full moon, 20:50 UT.
Moon near Saturn tonight.

December 3 Moon near Jupiter this morning.

4 Mercury at superior conjunction

7 Last quarter moon, 19:53 UT.

14 New moon, 20:48 UT.
Geminid meteor shower peaks this morning (9h UT).
Annular solar eclipse begins 18:03 UT — for details see page 38.

20 Moon near Mars this evening.

21 Winter begins (winter solstice, 19:23 UT).

22 First quarter moon, 20:57 UT.

28 Moon nears Saturn this evening.

30 Full moon, 10:42 UT.
Moon near Jupiter this evening.

GLOSSARY

Altitude The angular distance, usually measured in degrees, of an object above the horizon.

Aphelion The point where an object in orbit around the sun is farthest from it.

Apogee The point where an object in orbit around the earth is farthest from it.

Aurora Regions of glowing gas in the upper atmosphere whose molecules are stimulated to emit light by collisions with streams of electrons.

Asteroid A rocky body less than 620 miles (1,000 km) across that orbits the sun; more accurately called a minor planet or a planetesimal.

Astrology An ancient system of beliefs that attempts to explain or predict human actions by the position and interaction of the sun, moon, and planets. It is not a science.

Asterism A noticeable pattern of stars, such as the Big Dipper or the Pleiades, that is part of a larger constellation.

Axis An imaginary line passing through the center of a body, such as a planet, around which that body spins.

Azimuth The angular distance, usually measured in degrees, of an object's direction along the horizon starting from north (0° or 360°) through east (90°), south (180°), and west (270°).

Binary star A system containing two or more stars in orbit around one another.

Bolide A fireball that breaks up during its passage through the atmosphere.

Comet A small body made of ice and rock that orbits the sun, usually much less than 62 miles (100 km) across. As it nears the sun it usually brightens and develops a gaseous halo (coma) and a tail of gas and dust. Most comets travel in very elongated orbits that keep them far from the inner solar system.

Conjunction The alignment of two celestial bodies that occurs when they share similar angles from the sun as measured along the ecliptic. This is also roughly when the bodies appear closest together in the sky. **Inferior conjunction:** That point in the motions of the planets Mercury and Venus at which they pass between Earth and the sun. **Superior conjunction:** That point in the motions of Mercury and Venus at which they appear in line with the sun on the far side of their orbits as viewed from Earth. **With the sun:** That point in the motions of the superior planets at which they appear in line with the sun as viewed from Earth.

Constellation One of eighty-eight regions into which astronomers divide the sky, based mainly on earlier divisions formed by historical and mythological figures of Greek and Roman tradition.

Double star Two stars that appear close to one another. They can be physically associated (a binary) or simply appear together from the point of view of an observer on earth.

Earthshine A blue-gray light seen during the moon's crescent phases on the portion not illuminated by the sun. The sunlit portion of the earth is the source.

Eclipse An event during which the moon passes either in front of the sun (solar eclipse) or into the shadow of the earth (lunar eclipse) as seen from some locations. **Total solar eclipse:** A solar eclipse in which the moon entirely covers the sun. **Annular eclipse:** A solar eclipse in which the moon covers all but a thin ring, or annulus, of the sun.

Ecliptic   The apparent yearly path of the sun through the sky. Since this apparent motion is actually a reflection of Earth's movement, the ecliptic also marks the plane of Earth's orbit. The moon and planets also roughly follow this path.

Equinox   The date of the year at which the sun's rays illuminate half the earth, from pole to pole; neither the North Pole nor the South Pole is angled into the sun. This phenomenon occurs on two days of the year, near March 21 and September 23. On these dates, the hours of daylight equal the hours of night (hence the name, meaning "equal night"). The March equinox is considered the first day of spring in the northern hemisphere; the September equinox the first day of fall.

Fireball   An extremely bright meteor, usually those brighter than magnitude -4.

Galaxy   A vast collection of billions of stars, gas, and dust held together by the gravity of its members. The galaxy in which the sun resides is called the Milky Way.

Gas giants   The planets Jupiter, Saturn, Uranus, and Neptune.

Inferior conjunction   See Conjunction.

Light-year   The distance traveled through space by a beam of light in one year. Light travels at 186,282 miles (299,792 km) per second, so a light-year is 5.88 trillion miles (9.5 trillion km), or 63,240 times Earth's distance from the sun.

Magnitude   A measure of the relative brightness of stars and other celestial objects. The brighter the object the lower its assigned magnitude.

Meteor   The streak of light caused by a solid body in orbit around the sun (a meteoroid) passing through the atmosphere; also called a "shooting star." A **meteorite** is a meteoroid that strikes the surface of a planet or moon.

Meteor shower   The appearance of many meteors within a few hours that seem to radiate from the same region of the sky. They occur when the earth passes through the dusty debris near a comet's orbit.

Milky Way   A faint band of light around the sky composed of vast numbers of stars too faint to see individually. Also, the name of the galaxy in which the sun resides.

Moon   A natural satellite orbiting a planet. Also, the name of Earth's natural satellite.

Nebula   A cloud of gas and dust, sometimes glowing from the light of nearby stars and sometimes a dark patch that blocks starlight.

Nova   A star which suddenly erupts, greatly increasing its brightness.

Nucleus   Of a comet, the solid ice-rock mixture at the center of a comet's gaseous head and tail. Of a spiral galaxy, the dense central portion made of older, redder stars.

Opposition   The point in a planet's orbit at which it appears opposite the sun in the sky. A planet at opposition is visible all night long. Because they orbit closer to the sun than Earth, Mercury and Venus never reach opposition.

Perigee   The point where an object in orbit around the earth is nearest to it.

Perihelion   The point where an object in orbit around the sun is nearest to it.

Phases   The cycle of varying shape in the sunlit portion of a planet or moon. The moon, Venus, and Mercury all show phases as seen from Earth.

Photosphere   The visible surface of the sun.

Planet   A body of substantial size held in orbit by the gravity of a star. A planet shines by reflecting the star's light.

Radar   Radio signals transmitted to and bounced back from an object. It stands for *RA*dio *D*etection *A*nd *R*anging.

Radiant   The point in the sky from which shower meteors seem to appear.

Radiation   Energy transmitted through space as waves or particles.

Retrograde motion   The apparent backward (westward) loop in a planet's motion across the sky. All planets display retrograde motion, but that of Mars is most striking.

Satellite   A natural or artificial body in orbit around a planet.

Scintillation   A tremulous effect of starlight — twinkling — caused by its passage through our turbulent atmosphere. Planets usually don't exhibit this effect.

Sidereal period   The time taken by a planet to complete one revolution around the sun (or for the moon to complete an orbit around the earth) as measured by reference to the background stars.

Solar wind   A stream of electrically charged particles from the sun.

Solstice The date of the year at which either the earth's North or South Pole is angled most directly toward the sun. This occurs on two days of the year, near June 21 and December 21. The June solstice, the longest day of the year in the northern hemisphere, is considered the first day of summer there; the sun makes its most northerly arc through the sky. The December solstice, the shortest day of the year, marks the start of northern winter; the sun then makes its most southerly arc through the sky.

Star A hot, glowing sphere of gas, usually one that emits energy from nuclear reactions in its core. The sun is a star.

Sunspot A magnetic disturbance on the sun. It is cooler than the surrounding area and, consequently, appears darker.

Superior conjunction See Conjunction.

Supernova An enormous stellar explosion that increases the brightness of a star by a factor of more than 100,000. Although the star itself is destroyed, a small portion of its central core may survive (a neutron star).

Synodic period The average time between successive returns of a planet or the moon to the same apparent position relative to the sun — for example, new moon to new moon, or opposition to opposition.

Terminator The edge of the sunlit portion of the moon or planets; the line between day and night.

Terrestrial planets Mercury, Venus, Earth, and Mars.

Tides Periodic changes in the shape of a planet, moon, or star caused by the gravity of a body near it.

Train A dimly visible path left in the sky by the passage of a meteor.

Transit The passage of a planet across the face of the sun. From Earth, only Mercury and Venus can transit.

Twinkling See Scintillation.

Variable star A star whose brightness changes.

White dwarf A collapsed object formed from a star that has exhausted its nuclear fuel. The sun will one day become a white dwarf.

Zenith The point directly overhead, 90° above the horizon.

Zodiac The band of twelve constellations straddling the ecliptic.

BIBLIOGRAPHY

History and Mythology

Allen, R. H.. *Star Names: Their Lore and Meaning.* New York: Dover Publications, Inc., 1963.

Aveni, Anthony, F. *Skywatchers of Ancient Mexico.* Austin, TX: University of Texas Press, 1980.

Brekke, A. and A. Egeland. *The Northern Light: From Mythology to Space Research.* Berlin: Springer-Verlag, 1983.

Cary, M., et al., eds. *The Oxford Classical Dictionary.* Oxford: Oxford University Press, 1968.

Clark, David H. and F. Richard Stephenson. *The Historical Supernovae.* Oxford: Pergamon Press, 1977.

Drake, Stillman. *Discoveries and Opinions of Galileo,* Garden City, NY: Doubleday and Co., Inc., 1957.

Dreyer, J. L. E. *A History of Astronomy from Thales to Kepler.* New York: Dover Publications, Inc., 1953.

Ferris, Timothy. *Coming of Age in the Milky Way.* New York: Doubleday, 1989.

Graves, Robert. *The Greek Myths.* New York: Penguin Books, 1960.

Hodson, F. R., ed. *The Place of Astronomy in the Ancient World.* London: Oxford University Press, 1974.

Krupp., E. C. *Beyond the Blue Horizon: Myths and Legends of the Sun, Moon, Stars, and Planets.* New York: Harper and Row 1991.

Krupp., E. C. *Echoes of the Ancient Skies: The Astronomy of Lost Civilizations.* New York: Harper and Row, 1983.

O'Neil, W. M. *Early Astronomy from Babylonia to Copernicus.* Sydney: Sydney University Press, 1986.

Sanderson, Richard. "The Night it Rained Fire," *Griffith Observer,* November 1984, 2-10.

Schafer, Edward H. *Facing the Void: T'ang Approaches to the Stars.* Berkeley: University of California Press, 1977.

Stephenson, F. Richard. "Guest Stars are Always Welcome," *Natural History,* August 1987, 72-76.

Stephenson, Richard and Kevin Yau. "Oriental Tales of Halley's Comet," *New Scientist,* September 27, 1984, 31-32.

Toulman, Stephen and June Goodfield. *The Fabric of the Heavens: The Development of Astronomy and Dynamics.* New York: Harper and Row, 1961.

Upton, Edward K. L. "The Leonids Were Dead, They Said," *Griffith Observer,* May 1977, 3-9.

Williamson, Ray A. *Living the Sky: The Cosmos of the American Indian.* Boston: Houghton Mifflin Co., 1984.

Science

Beatty, J. Kelly and Andrew Chaikin, eds. *The New Solar System,* third edition. Cambridge, MA: Sky Publishing Corp., 1990.

Carr, Michael H. *The Surface of Mars.* New Haven: Yale University Press, 1981.

Cook, Allan F. "A Working List of Meteor Streams," in *Evolutionary and Physical Properties of Meteor Streams,* NASA SP-319, 183-191, 1973.

Dodd, Robert T. *Thunderstones and Shooting Stars: The Meaning of Meteorites.* Cambridge, MA: Harvard University Press, 1986.

Espenak, Fred. *Fifty Year Canon of Solar Eclipses, 1986-2035.* NASA Reference Publication 1178 Revised, July 1987. Cambridge, MA.: Sky Publishing Corp., 1987.

Espenak, Fred. *Fifty Year Canon of Lunar Eclipses, 1986-2035.* NASA Reference Publication 1216, March 1989. Cambridge, MA: Sky Publishing Corp., 1989.

Halliday, Ian, Alan T. Blackwell, and Arthur A. Griffin. "The Frequency of Meteorite Falls on the Earth," *Science,* volume 223, 1405-1407, 1984.

Hartmann, William K. *Astronomy: The Cosmic Journey,* fourth edition. Belmont, CA: Wadworth Publishing Co., 1989.

Helfand, David. "Bang: The Supernova of 1987," *Physics Today,* August 1987, 25-32.

Kohlhase, Charles, ed. *The Voyager Neptune Travel Guide,* Pasadena, CA: Jet Propulsion Laboratory Publication 89-24.

Hunt, Garry and Patrick Moore. *Jupiter.* London: Mitchell Beasley, 1982.

Hunt, Garry and Patrick Moore. *Saturn.* London: Mitchell Beasley, 1981.

Kronk, Gary. *Meteor Showers: A Descriptive Catalog.* Hillside, New Jersey: Enslow Publishers, 1988.

Marsden, Brian G. and Gareth V. Williams. *Catalogue of Cometary Orbits, 1992.* Cambridge, MA: Minor Planet Center, Smithsonian Astrophysical Observatory, 1992.

Wilson, Andrew. *Solar System Log.* London: Jane's Publishing Company Ltd., 1987.

Yeomans, D. K. "Comet Tempel-Tuttle and the Leonid Meteors," *Icarus,* volume 47, 492-499, 1981.

Zirker, J. B. *Total Eclipses of the Sun.* New York: Van Nostrand Reinhold Co., Inc., 1984.

Observing Guides

Burnham, Robert, Jr. *Burnham's Celestial Handbook.* New York: Dover Publications, Inc., 1978.

Meeus, Jean. *Astronomical Tables of the Sun, Moon, and Planets.* Richmond, VA: Willmann-Bell, 1983.

Ottewell, Guy. *The Astronomical Companion.* Greenville, SC: Furman University, 1983.

Ottewell, Guy. *The Astronomical Calendar.* Greenville, SC: Furman University (annual).

Victor, Robert C., et al. *Sky Calendar.* East Lansing, MI: Abrams Planetarium, Michigan State University (monthly).

INDEX

Sky Log

Sky Log

Sky Log

Sky Log

Sky Log

Sky Log

Sky Log

Sky Log

Sky Log